Report of Investigations No. 145

Origin of Silver-Copper-Lead Deposits in Red-Bed Sequences of Trans-Pecos Texas: Tertiary Mineralization in Precambrian, Permian, and Cretaceous Sandstones

Jonathan G. Price
Christopher D. Henry
Allan R. Standen
Jan S. Posey

Research for this project was funded by the U.S. Department of the Interior, originally through the Office of Surface Mining and later through the U.S. Bureau of Mines, under grant number G114037.

1985

Bureau of Economic Geology • W. L. Fisher, Director
Texas Mining and Mineral Resources Research Institute
The University of Texas at Austin • Austin, Texas 78713

CONTENTS

ILLUSTRATIONS

Figures

Tables

Appendix Tables

ABSTRACT

Silver deposits occur in Precambrian, Permian, and Cretaceous red-bed sequences near Van Horn, Texas. These deposits are geochemically similar and contain economically important quantities of silver, copper, and lead, as well as anomalously high amounts of arsenic, zinc, cadmium, and molybdenum. Gold is not enriched. Primary minerals include chalcopyrite, tennantite-tetrahedrite, bornite, galena, sphalerite, acanthite, pyrite, marcasite, barite, and calcite.

The deposits are dominantly steeply dipping veins. Strata-bound occurrences are near veins or closely spaced fractures. Structural evidence, including orientations of veins, relative ages of fractures, and relationships to major tectonic events in the region, suggests that the most likely time of mineralization was during late Basin and Range extensional deformation. Ore deposition probably occurred at least 18 m.y. after the period of voluminous silicic volcanism (38 to 28 m.y.a.) in the Trans-Pecos region.

Other features indicate that, in contradiction to a hypothesis suggested by previous workers, the deposits did not form as a result of middle Tertiary magmatism. (1) Centers of igneous activity were distant from the sites of mineralization. (2) Potentially reactive limestones above and below the ore zones in red beds are generally unmineralized. (3) Zones of argillic, phyllic, and propylitic alteration typical of igneous-hydrothermal veins are absent. (4) Characteristic igneous-hydrothermal gangue minerals such as quartz and fluorite are rare or absent. (5) Homogenization temperatures of fluid inclusions in barite and calcite suggest formation temperatures in the range of 120°C to 170°C, that is, lower than temperatures typical in copper-lead-zinc-bearing igneous-hydrothermal veins. These temperatures are higher than those usually attributed to strata-bound, red-bed copper deposits. The elevated formation temperatures are the result of high heat flow in the Basin and Range province of Texas at the time of mineralization; they are not the result of igneous activity.

The hypothesis developed in this study on the origin of the veins involves a rise of moderate-temperature, moderately saline hydrothermal fluids along Basin and Range fractures and precipitation of metal sulfides in response to mixing with shallow ground water. This hypothesis has implications for exploration of additional deposits in Trans-Pecos Texas and elsewhere.

Keywords: Texas, Trans-Pecos, economic geology, metals, silver, copper, lead, red beds

INTRODUCTION

Purpose of Study

Silver-copper-lead deposits in red-bed sequences near Van Horn in Trans-Pecos Texas (fig. 1) have yielded large quantities of ore from Precambrian, Permian, and to a lesser extent Cretaceous sandstones (table 1). Recent mining in the Precambrian Belt Series of Montana and Idaho (Dayton, 1983; Lange and Sherry, 1983) has demonstrated the feasibility of large-scale mining of low-grade, strata-bound silver-copper deposits in red-bed sequences. Although some of the Trans-Pecos ores are strata-bound, major production has come from nearly vertical veins that crosscut sedimentary layers. The purpose of this study was to determine the origin of these veins.

Questions addressed include

(1) whether the mineral deposits in host rocks of different age formed at the same or different times;

(2) how the deposits relate to major tectonic events in the region, including Precambrian and Paleozoic deformations, Late Cretaceous to Eocene folding and thrusting, Oligocene igneous activity, and younger extensional faulting in the Basin and Range province;

(3) what the physical and chemical conditions of ore deposition were; and

(4) how understanding the origin of the known deposits applies to exploration for additional veins and strata-bound ores.

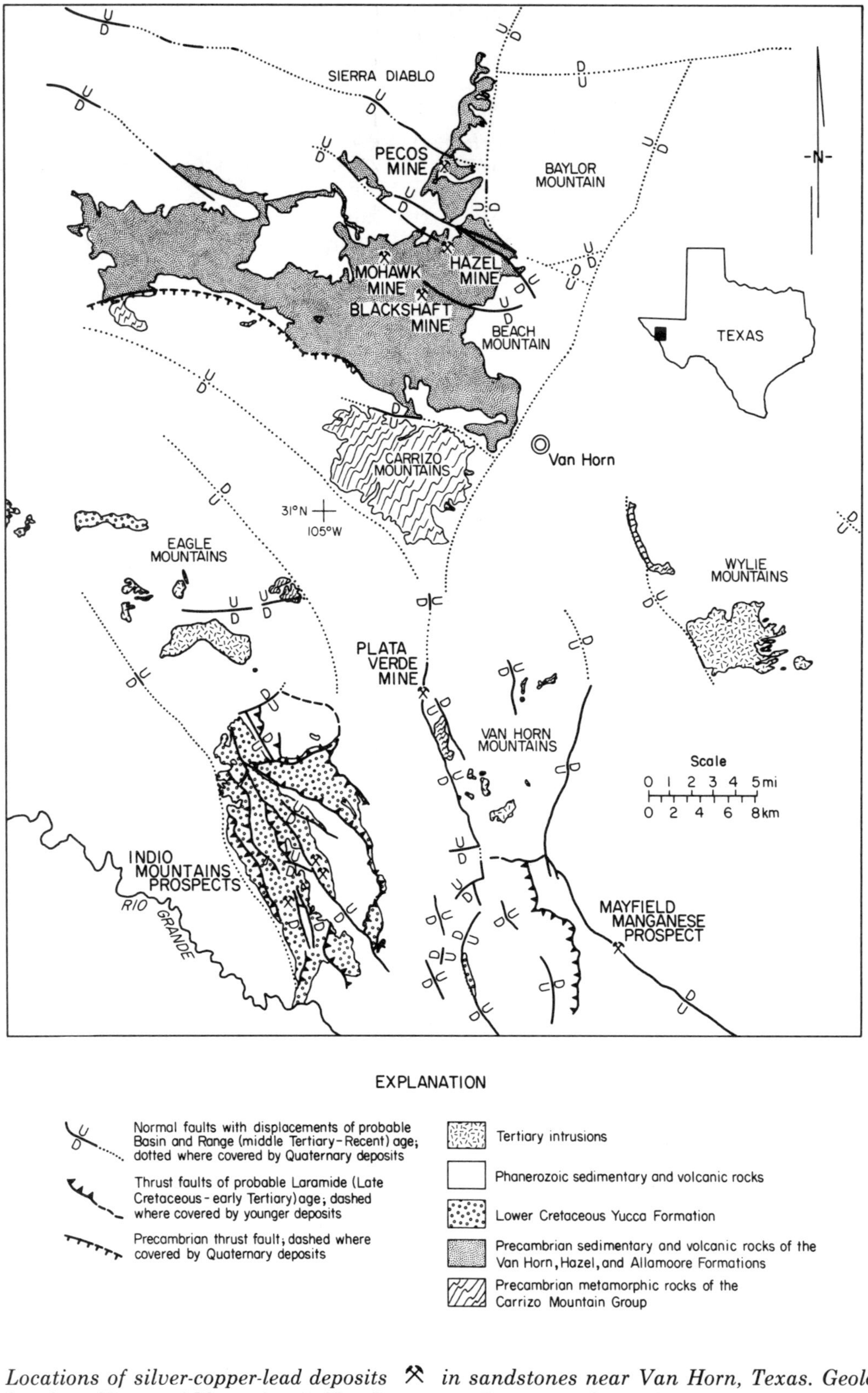

FIGURE 1. Locations of silver-copper-lead deposits ⚒ in sandstones near Van Horn, Texas. Geologic features shown are taken from King and Flawn (1953), Hay-Roe (1957), Twiss (1959b), Underwood (1963), King (1965), Barnes (1968, 1979), and Henry (1979). Orebodies in the Van Horn - Allamoore district (Hazel, Pecos, Mohawk, Blackshaft mines) occur chiefly in the Precambrian Hazel Formation in the southern Sierra Diablo area. The Powwow Member of the Permian Hueco Limestone, which unconformably overlies Precambrian rocks in most of the area, hosts ore at the Plata Verde mine. Mineralization in the Indio Mountains is restricted to the Cretaceous Yucca Formation.

TABLE 1. Estimated total production, silver grade, copper-to-silver ratio, and silver-to-gold ratio of silver-copper-lead deposits in red-bed sequences in Trans-Pecos Texas.[1]

Host rock	Total production (tons)	Silver grade (oz/ton)	Cu/Ag (by weight)	Ag/Au (by weight)
Precambrian Hazel Formation	110,000	36	5	>160,000
Permian Powwow Member of Hueco Limestone	16,000	17	6	87,000
Cretaceous Yucca Formation	<200	1.2	200	>8,400

[1]Data from Flawn (1952), King and Flawn (1953), Wallace (1972), Price (1982), and analyses of high-grade samples from mine workings and dumps.

Scope of Study

This study discusses mineral deposits in red beds of the Trans-Pecos region. Additional silver and copper deposits in the region are only briefly mentioned for comparison. King and Flawn (1953) subdivided the mineral deposits in Precambrian rocks of the Van Horn - Allamoore district (fig. 2) into four main groups.

(1) The **Hazel-type deposits** (including the Hazel, Mohawk, Pecos, and Hackberry mines and the Marvin - Judson, Diablo, and Eureka prospects), which are the subject of this study, were the most silver-productive veins. All the Hazel-type deposits occur along steeply dipping fractures in the Precambrian Hazel Formation (fig. 3).

(2) The **Blackshaft-type deposits** (including the Blackshaft, Saint Elmo, and Sancho Panza mines) produced less silver than did the Hazel-type deposits but yielded similar quantities of copper (table 2). These deposits are not obviously vein related but appear to be confined to a bed that comprises stromatolitic dolomite, limestone, tuffaceous sandstone, shale, local basalt, and chloritic fault gouge. On the basis of similarities in lithology, King and Flawn (1953) interpreted this bed as a thin slice of the Allamoore Formation that was thrust on top of and conformably overlain by Hazel Formation red beds. Reid (1974), however, concluded that this sequence of Allamoore-like rocks is a bed within the Hazel Formation. Unlike the Hazel-type veins, the Blackshaft-type deposits are stratiform (fig. 4) and geochemically distinct. Although copper grades in the two types are similar, the Hazel ores are much richer in silver than are the Blackshaft ores (table 3). Whereas the Hazel-type veins generally cut nearly horizontal red beds, the rocks to the south near the Blackshaft-type deposits were folded and somewhat metamorphosed (graphite is locally present) during Precambrian deformation. More work is needed to decipher possible syngenetic and epigenetic or metamorphic features of the Blackshaft deposits.

TABLE 2. Estimated ore, copper, and silver production from mines in the Van Horn - Allamoore district.[1]

Mine	Ore (tons)	Copper (lb)	Silver (oz)
Hazel	110,000	1,500,000	4,000,000
Blackshaft	13,000	740,000	4,000
Sancho Panza	8,500	400,000	7,000
Hackberry	610	8,600	20,000
Saint Elmo[2]	350	16,000	290
Mohawk[3]	300	6,600	2,600
Pecos[3]	100	2,200	880
Total for district	130,000	2,700,000	4,000,000

[1]Modified from King and Flawn (1953).
[2]Estimated grades of copper (2.3%) and silver (0.82 oz/ton) are typical of the similar deposits nearby at the Sancho Panza mine.
[3]Estimated grades of copper (1.1%) and silver (8.8 oz/ton) are typical of 1927 to 1947 production from similar deposits at the Hazel mine.

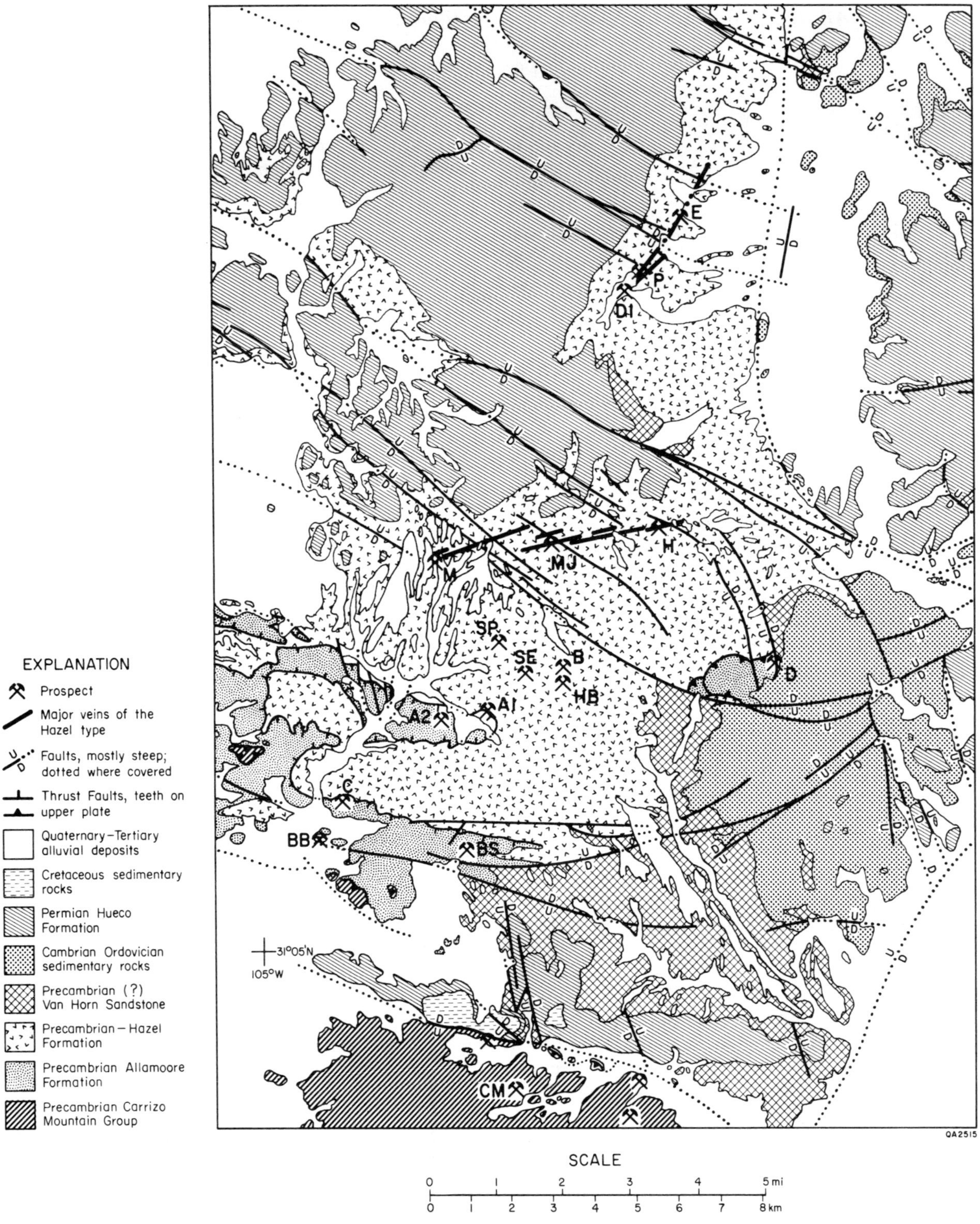

FIGURE 2. Geology and locations of silver-copper-lead deposits in Precambrian rocks of the Van Horn -Allamoore district. See figure 1 for location. Hazel-type veins occur at the Hazel (H), Mohawk (M), Pecos (P), and Hackberry (HB) mines and at the Marvin - Judson (MJ), Eureka (E), and Diablo (DI) prospects. Blackshaft-type strata-bound deposits occur at the Blackshaft (B), Saint Elmo (SE), and Sancho Panza (SP) mines. Other occurrences of silver-copper-lead deposits include the Dallas (D), Anaconda No. 1 (A1) and No. 2 (A2), Bluebird (BB), Buck Springs (BS), and Cooper Hill (C) prospects. Copper is also associated with metamorphic quartz veins in the Carrizo Mountains (CM). Geologic features are from King (1965).

FIGURE 3. Glory hole east of the East shaft, Hazel mine. Hazel-type veins follow vertical fractures in sandstones and coarse siltstones. Bush (upper left) is approximately 3 ft (1 m) high.

(3) The **Dallas prospect** (fig. 2) is unusual because it occurs along a fault that King and Flawn (1953) determined was active before Ordovician time. The host rocks of the copper mineralization at the Dallas prospect are the Precambrian Allamoore Formation and Van Horn Sandstone.

(4) Scattered **Allamoore Formation deposits** (Anaconda No. 1 and No. 2, Bluebird, Buck Springs, and Cooper Hill prospects) are mostly small copper veins having no known production. Copper-bearing quartz veins of probable metamorphic origin occur in various types of metasedimentary and metaigneous rocks in the Precambrian Carrizo Mountain Group.

Of the four groups of deposits in Precambrian rocks, only the Hazel-type veins are found exclusively in red beds. In Trans-Pecos Texas, the other red-bed-hosted deposits are (1) the strata-bound secondary ores at the Plata Verde mine in the clastic Powwow Member of the Permian Hueco Limestone and (2) the steeply dipping veins (fig. 5) and locally strata-bound occurrences in the Cretaceous Yucca Formation of the Indio Mountains. The Hazel-type veins, Plata Verde ores, and Indio Mountains prospects, which are the subjects of the report, are markedly similar in structure and geochemistry.

FIGURE 4. Stratiform ore at the Sancho Panza mine. Gently dipping to the left, red Hazel sandstone overlies copper-rich tuff and stromatolite.

TABLE 3. Average copper grade, silver grade, and copper-to-silver ratio, based on production records of mines in the Van Horn - Allamoore district.[1]

Deposit type	Mine	Amount of ore for which complete records are known (tons)	Copper grade (%)	Silver grade (oz/ton)	Cu/Ag (by weight)
Hazel	Hazel: 1891	600	3.5	589	2
	1927-1947	36,876	1.1	8.8	36
	Hackberry	610	0.7	32	6
	Mohawk	4	3.5	20	52
Blackshaft	Blackshaft	7,900	3.0	0.37	2,400
	Sancho Panza	5,400	2.3	0.82	820

[1]Data from King and Flawn (1953).

FIGURE 5. Nearly vertical vein (dark area) cutting Cretaceous Yucca Formation conglomerate at the Rossman prospect, Indio Mountains. Note hammer for scale.

Previous Work

Most of the earlier geological investigations of silver-copper-lead deposits in Texas discussed the deposits in the Precambrian Hazel Formation northwest of Van Horn (fig. 1). Von Streeruwitz (1890, 1891, 1892a, 1892b, 1893) examined the deeper workings at the Hazel mine, the most productive of all the deposits. In describing the geology, mineralogy, workings, and production at the Hazel mine, Flawn (1952) summarized the observations made by von Streeruwitz, Richardson (1914), and Sample and Gould (1945). Descriptions of other deposits in the Van Horn - Allamoore district were incorporated with new observations by King and Flawn (1953) in their study of Precambrian rocks in the area. King (1965) later expanded his investigations to include the entire Sierra Diablo region. Reid (1974) studied the sedimentology of the Hazel Formation.

Twiss (1959a, 1959b) mapped and described the geology of the Van Horn Mountains, the site of the Plata Verde mine in the Permian Hueco Formation (fig. 1). Brief comments on the mine by Sellards and Evans (1946) were included in a comprehensive study by Price (1982).

Adams (1953), Allday (1953), Bostwick (1953), and Underwood (1962, 1963, 1980) described the geology of the Indio Mountains, where small silver-copper-lead veins occur in the Cretaceous Yucca Formation (fig. 1). Flawn (1950) reported on mineral production in the area. Wallace (1972) used Underwood's mapping as a base for more detailed descriptions of the Indio Mountains prospects. Wallace and Shannon (1975) summarized the work by Wallace (1972). Price and others (1983a) presented new structural observations, which are included in this study. In summaries of previous work, Evans (1975) discussed silver and gold ores throughout Texas. Price and others (1983b) compiled data on all known Trans-Pecos mineral deposits. Hypotheses discussed in the present study were first outlined by Price (1983).

Previous workers who commented on the ages and origins of the vein deposits (for example, King and Flawn [1953] and Wallace [1972]) suggested that the veins formed in response to Tertiary igneous activity. Although hydrothermal veins and other types of ores related to igneous activity are common in the Trans-Pecos region, evidence presented in this study suggests that silver-copper-lead mineralization in red-bed sequences is younger than and not directly related to Tertiary igneous activity.

History of Mining

The largest vein in the Precambrian Hazel Formation is located at the Hazel mine (fig. 6), and the vein accounts for most of the total silver production (tables 1 and 2). The history and workings of the mine were discussed in detail by Flawn (1952) and King and Flawn (1953). The outcropping vein was discovered in 1856 but because of Indian raids was not developed substantially until 1880. A total of approximately 4,000,000 oz (120,000 kg) of silver was produced from the Hazel mine, mostly between 1880 and 1896. Deepest workings, at the East shaft, were 746 ft (227 m) below the surface. Flawn (1952) reported that the host rock at the bottom of the shaft was probably conglomerate in the lower part of the Hazel Formation. By 1900, flooding of the lower levels had prevented deep mining. The vertical extent of high-grade silver ore is unknown. Lead content of the ore apparently was economically insignificant; no records of lead production exist. Sporadic production of copper and silver continued at the Hazel mine until 1947. Other mines in the Van Horn - Allamoore district produced lead, copper, and silver intermittently into the 1960's.

Between 1934 and 1943, the Plata Verde mine yielded approximately 279,000 oz (8,700 kg) of silver plus substantial quantities of copper and lead (Price, 1982). Only oxide ore was produced from the workings, the deepest of which was 160 ft (49 m) below the surface. Exploratory trenching and shallow diamond drilling in the early 1980's have indicated additional oxide-ore reserves at the mine.

Prospecting has been intermittent in the remote Indio Mountains area from the early 1900's to the present. None of the numerous deposits in the Indio Mountains was large enough to sustain production. Judging from the sizes of exposed workings and dumps, it is likely that only a few hundred tons of ore were removed from the area.

FIGURE 6. Headframe at the East shaft, Hazel mine. Limestone cliffs of the Permian Hueco Limestone overlie the Precambrian Hazel Formation, host of the silver-copper ore.

Methods of Investigation

The approach taken to determine the origin of silver-copper-lead deposits in red-bed sequences involved field and laboratory studies. Geologic mapping by King and Flawn (1953) in the Precambrian rocks (fig. 2) and by Price (1982) at the Plata Verde mine (fig. 7) provides an adequate base for this study, but more detailed mapping was required in the Indio Mountains. Geologic relations complicated by Late Cretaceous to Eocene thrusting and folding and by middle to late Tertiary Basin and Range faulting in an area encompassing the Indio Mountains prospects were mapped on a topographic base at a scale of 1:6,000. This map, reduced and without the topographic base, appears here as figure 8.

Orientations of fractures and veins were measured in and near each of the three main groups of ore deposits. Fracture measurements

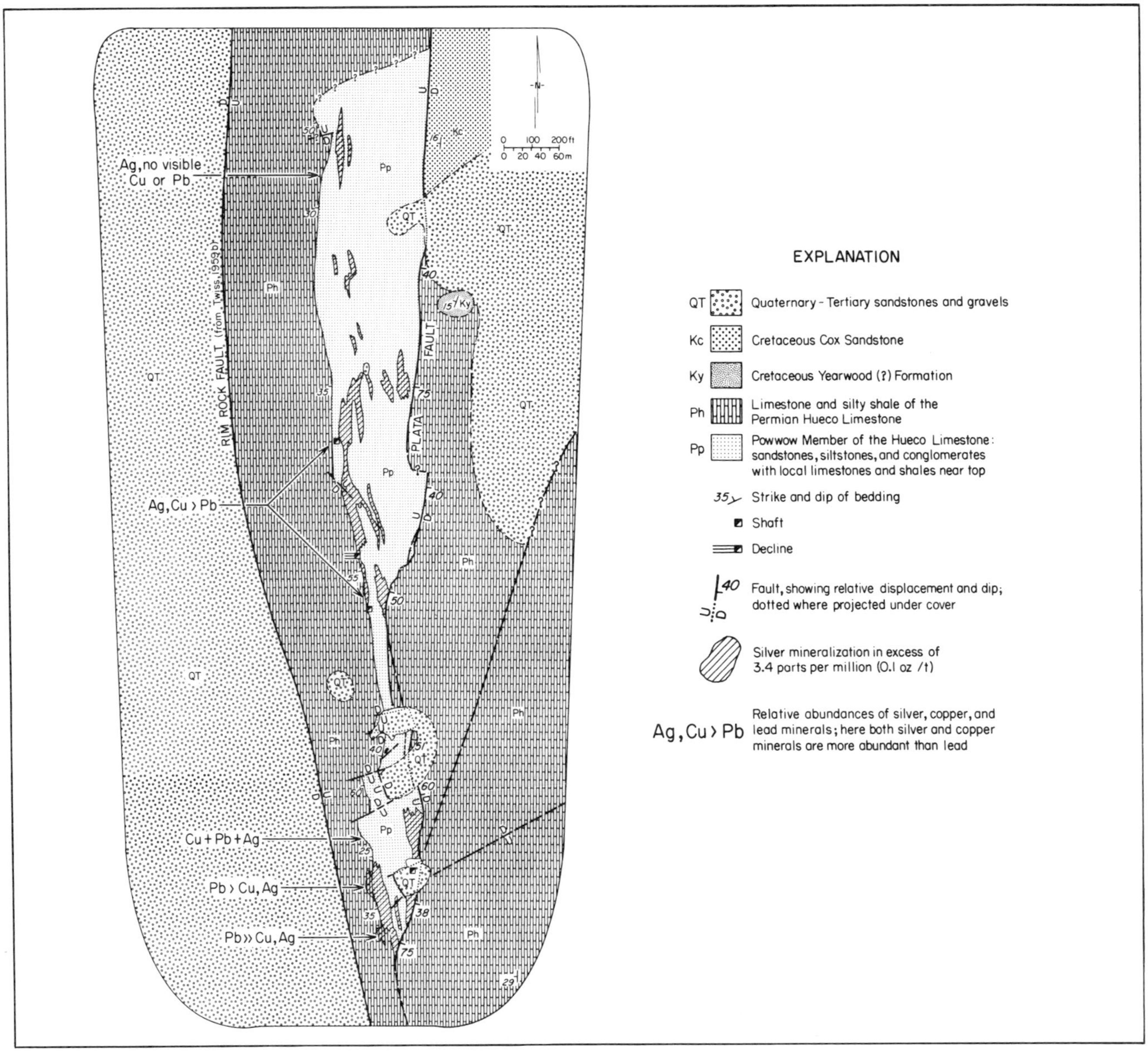

FIGURE 7. Simplified geologic map of the Plata Verde mine, Trans-Pecos Texas. Modified from Price (1982). See figure 1 for location. A north-trending Basin and Range horst is the dominant structure. Except for minor amounts of silver in siltstones and shales in the lower part of the limestone member of the Permian Hueco Limestone, silver is restricted to the westward-dipping sandstones of the Powwow Member. North-south zonation of silver-copper-lead minerals correlates with a southward increase in reduction within the Powwow sandstones and with more numerous faults in the southern part of the mine area.

were also made in Permian rocks overlying the Precambrian Hazel Formation at the base of the Sierra Diablo escarpment near the Hazel and Pecos mines and the Marvin - Judson prospect (fig. 2). Well-preserved slickenlines in the Yucca Formation were measured in the Indio Mountains.

Selected samples collected during field studies were analyzed geochemically to determine relative elemental enrichments and depletions in ores versus unmineralized rocks. Samples were examined using reflected- and transmitted-light microscopy and X-ray diffraction to determine mineralogy and to establish textural relationships. Fluid-inclusion studies revealed information about the temperatures and salinities of the ore-forming fluids. Carbon and sulfur isotope analyses yielded data on the compositions and origins of the ore-forming fluids.

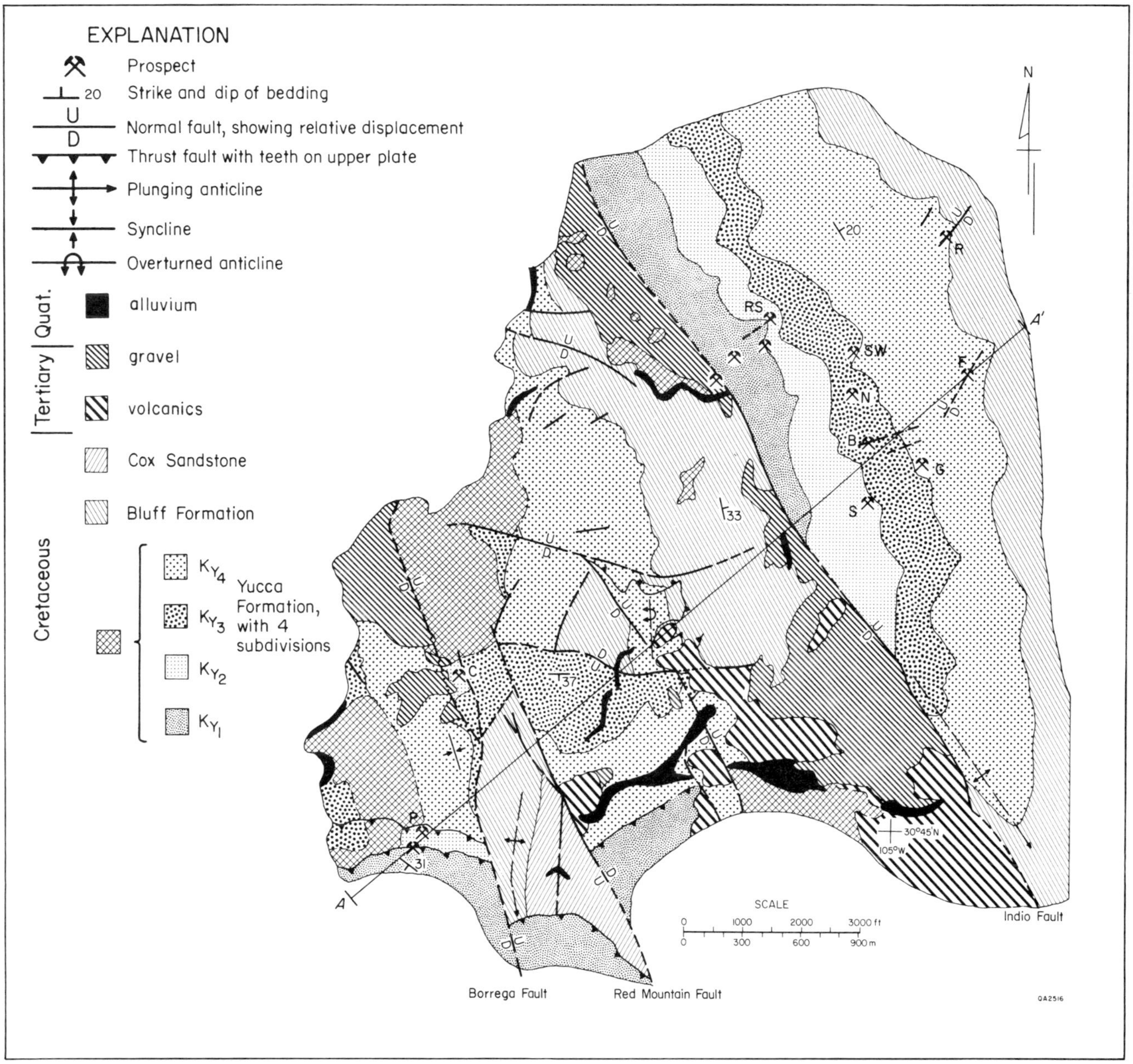

FIGURE 8. Geologic map of the central Indio Mountains. Modified from Allday (1953), Underwood (1963), and Wallace (1972). See figure 1 for location. Silver-copper-lead deposits, which are restricted to the Yucca Formation, include the Black Diamond (B) and Purple Sage (P) mines and the Carpenter (C), Finlay (F), galena-barite (G), Road (R), and Rossman (RS) prospects and the North pit (N), South pit (S), and Southwestern workings (SW). Thrust faulting and folding occurred during Late Cretaceous to Eocene Laramide deformation. Normal faulting occurred during late Oligocene to Recent Basin and Range extension.

Rocks ranging in age from Precambrian to Recent are exposed in the Van Horn area (fig. 9). Unconformities recognized in the stratigraphic column correspond to major tectonic events: late Precambrian deformations involving folding, thrusting, and metamorphism; tilting, uplift, and erosion associated with Late Pennsylvanian to Early Permian (Ouachita-Marathon) folding and thrusting 110 mi (170 km) to the southeast; Late Cretaceous to Eocene (Laramide) folding and thrusting; Eocene and Oligocene igneous activity; and Oligocene to Recent Basin and Range extension.

Silver-copper-lead deposits occur in three red-bed sequences (mostly sandstones): (1) the Precambrian Hazel Formation in the Van Horn - Allamoore district (fig. 2), (2) the Powwow Member of the Permian Hueco Limestone (fig. 7), and (3) the Cretaceous Yucca Formation in the Indio Mountains (fig. 8). Although stratigraphically confined, the ores are structurally controlled and probably formed much later than their enclosing host rocks. The following sections on stratigraphy and structure and tectonics of the Van Horn area provide a basis for understanding the origin of the ore deposits. It will be shown that timing of mineralization is linked to the tectonic history of the region and that stratigraphic controls on the location of ores are minor.

Stratigraphy

Precambrian Rocks

Two major packages of upper Precambrian rocks in the Van Horn area are separated by the Streeruwitz overthrust (King and Flawn, 1953): the Carrizo Mountain Group to the south, above the thrust, and the Allamoore and Hazel Formations to the north, below the thrust (figs. 1 and 2). Relative ages of the Carrizo Mountain Group and the Allamoore - Hazel sequence have not been determined because the two packages of rocks occur only in thrust contact.

The Carrizo Mountain Group (fig. 9) consists of quartzite, meta-arkose, slate, muscovite schist, biotite schist, metarhyolite, metabasalt (greenstone and amphibolite), pegmatite, and minor amounts of granodiorite and carbonate metasedimentary rocks. The metabasalts were probably sills (King and Flawn, 1953) and the metarhyolites were dominantly ash-flow tuffs (Rudnick, 1983). Total thickness of the sequence is approximately 19,000 ft (5,800 m) (King and Flawn, 1953). Condie (1982) and Rudnick (1983) suggested that the Carrizo Mountain Group rocks were deposited in a back-arc basin along what at the time of deposition was the continental margin. Using whole-rock Rb/Sr isochrons, Denison (1980) interpreted the age of deposition to be between 1,200 and 1,300 m.y.

North of the Streeruwitz Fault, the oldest exposed unit is the Allamoore Formation, which consists of cherty limestone and dolomite, talcose phyllite, tuff, and basalt flows and sills (King and Flawn, 1953). The Allamoore Formation is overlain by conglomerates, sandstones, and siltstones of the Hazel Formation. The upper part of the Allamoore Formation may interfinger with the Hazel Formation (W. B. Bourbon, personal communication, 1983). Reid (1974) recognized the Blackshaft stratigraphic horizon, which includes stromatolitic dolomite, limestone, tuffaceous sandstone, shale, and basalt, to be a zone of Allamoore-like lithologies within the Hazel Formation. Because folding was locally intense, thicknesses of the Allamoore and Hazel Formations are difficult to determine. King (1965) estimated the Allamoore Formation to be at least 2,500 ft (760 m) thick and the Hazel Formation to be approximately 5,000 ft (1,500 m) thick near the silver-copper-lead deposits. Unpublished seismic data suggest the presence of approximately 10,000 ft (3,000 m) of nearly horizontal Allamoore-Hazel rocks a few miles west of the Van Horn - Allamoore district.

Reid (1974) postulated that the Hazel red beds were deposited in alluvial fans that were supplied with sediments derived dominantly from erosion of the Allamoore Formation in a source area to the south. Although clasts of Allamoore lithologies are abundant in the Hazel Formation, the Allamoore Formation was probably not the source of the detrital microcline, muscovite, and tourmaline. These components, along with much of the quartz, may have been derived from the Carrizo Mountain Group or older basement rocks. Condie (1982) suggested that the Allamoore - Hazel sequence was deposited in a back-arc basin. In correlating data on basement rocks of Texas and adjacent parts of New Mexico, J. R. Garrison, Jr. (personal communication, 1984) included the Allamoore - Hazel in the Swisher - De Baca (continental) rift terrane, with which rhyolitic and basaltic magmatism is associated. Garrison suggested that the Swisher - De Baca terrane may be older than the Llano gneiss terrane (1,100 m.y.)

SYSTEM/ERA	STRATIGRAPHIC UNIT	THICKNESS m	COLUMNAR SECTION	DESCRIPTION
Quaternary-Tertiary	Recent to Miocene alluvium and bolson fill	0 to 335+		Gravel, sand, silt, and clay
Tertiary	Eocene and Oligocene igneous and volcaniclastic rocks			Silicic ash-flow tuffs, air-fall tuffs, and volcaniclastic sediments; intermediate and mafic lava flows; intermediate and silicic intrusions; generally no igneous rocks in immediate vicinity of silver-copper-lead deposits
Cretaceous	Chispa Summit Formation	260 to 610+		Gypsiferous, calcareous shale, flaggy limestone, siltstone, sandstone, and minor amounts of coal
	Buda Limestone	41 to 73		Limestone
	Eagle Mountains Sandstone	9 to 40		Sandstone, limestone, and shale
	Loma Plata Limestone	207 to 668		Limestone and marl with interbedded shale at base
	Benevides Formation	17 to 49		Siltstone, sandstone, and local limestone
	Finlay Limestone	53 to 243		Limestone with thin beds of shale, siltstone, and sandstone
	Cox Sandstone	270 to 529		Sandstone with some shale, limestone, and conglomerate
	Bluff/Yearwood Formation	0 to 457		Limestone and sandstone
	Yucca Formation ⚒	0 to 630+	Y4 Y3 Y2 Y1	In Indio Mtns, informally subdivided into members: Y1 (conglomerate, 150+m), Y2 (conglomerate with some sandstone, 110m), Y3 (sandstone with some conglomerate, 120m) and Y4 (sandstone with 2.5m limestone near top, 250m). Sandstones and conglomerates are red to white
Permian	? Hueco Limestone limestone member	100 to 400		In Van Horn Mtns., limestone with interlayered greenish-to bluish-gray shale and siltstone near base
	Powwow Member ⚒	0 to 82+		At Plata Verde Mine, coarse, red sandstone with subordinate red siltstone, greenish-gray sandstone and siltstone, and orange conglomerate and sandstone; interlayered limestone, shale, siltstone, and sandstone near top, 0 to 21m
Ordovician	Montoya Dolomite	70 to 140		Cherty dolomite, dolomite, and sandstone
	El Paso Limestone	350		Dolomitic limestone, marly limestone, and calcareous sandstone
Cambrian-Ordovician	Bliss Sandstone	30 to 49		Sandstone
Precambrian (?)	Van Horn Sandstone	0 to 210		Red sandstone and conglomerate
Precambrian	Hazel Formation ⚒ ?	1,500 (?)		Red-colored fine-grained sandstone, coarse-grained siltstone, and conglomerate; local horizon of stromatolitic dolomite, limestone, tuffaceous sandstone, shale, and basalt
	Allamoore Formation ?	760 +		Cherty limestone and dolomite, talcous phyllite, tuff, and basalt
	Carizzo Mountain Group	5,800		Quartzite, meta-arkose, slate, mica schist, metarhyolite, metabasalt, and pegmatite

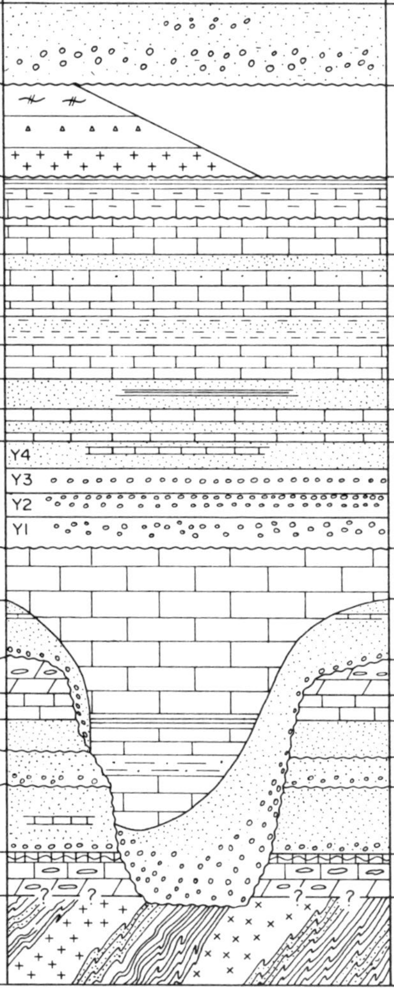

QA 2517

FIGURE 9. Stratigraphic column, Van Horn area. Modified from Twiss (1959a), Underwood (1963), and King (1965). Silver-copper-lead deposits ⚒ occur in the Precambrian Hazel Formation, the Powwow Member of the Permian Hueco Limestone, and the Cretaceous Yucca Formation. Wavy lines indicate unconformities.

with which he correlated the Carrizo Mountain Group.

The Van Horn Sandstone unconformably overlies rocks of the Hazel and Allamoore Formations and of the Carrizo Mountain Group. It in turn is unconformably overlain by the Cambrian through Ordovician Bliss Formation. The exact age of the Van Horn Sandstone is unknown, but Denison (1980) argued that it is probably Precambrian. The formation is composed of sandstone and conglomerate derived from the north and deposited in alluvial fans (McGowen and Groat, 1971). The dominant clast lithologies are granites and rhyolites similar or identical to approximately 1,000-m.y.-old rocks (Denison and Hetherington, 1969) exposed today 50 mi (80 km) to the northwest at Pump Station Hills and farther west in the Hueco and Franklin Mountains.

Paleozoic Rocks

Cambrian and Ordovician sedimentary rocks (fig. 9) crop out east of the Van Horn - Allamoore district (fig. 2). Silurian through Pennsylvanian sedimentary rocks may have been deposited and eroded before deposition of the Lower Permian Hueco Limestone. In places, the Hueco Limestone rests on each of the older formations.

The Hueco Limestone is subdivided into two members: (1) the lower Powwow Member, composed of fluvial conglomerates, sandstones, and siltstones and minor amounts of limestone and shale near the top and (2) the upper limestone member. In the Van Horn - Allamoore district, Powwow conglomerates capped by massive Hueco limestones unconformably overlie the Hazel Formation (figs. 2 and 6). Powwow sediments, which Hay-Roe (1957) and Twiss (1959a) suggested were deposited in alluvial fans, were derived from local sources (Price, 1982). In the Van Horn Mountains, the Powwow Member is composed dominantly of quartz, muscovite, and microcline derived from the Carrizo Mountain Group, upon which the Powwow Member rests. The thickest reported section of the Powwow Member, which is at least 270 ft (82 m) thick, occurs at the Plata Verde mine. Younger Permian rocks occur on the flanks of the Delaware Basin, a depositional center northeast of Van Horn that contains evaporites and clastic and carbonate rocks. Near Van Horn, Lower Cretaceous rocks unconformably overlie the Hueco Limestone.

The Sierra Diablo north of Van Horn was part of a generally high area, the Diablo Platform, throughout the Paleozoic. Lower and upper Paleozoic rocks thin across the platform, which remained a relatively stable and structurally elevated area well into Mesozoic time.

Mesozoic Rocks

Triassic and Jurassic rocks are absent in the Van Horn area but may have been deposited in the Chihuahua trough, a deep, evaporite-bearing sedimentary basin in Mexico near the Texas border. Cretaceous strata thicken southwestward in the Van Horn area, indicating the presence of the Chihuahua trough. The approximately 10,700-ft (3,300-m) section of Cretaceous rocks in the Indio Mountains (Underwood, 1963) is closer to the axis of the trough than is the 3,600-ft (1,100-m) section in the Van Horn Mountains (Twiss, 1959a). Cretaceous rocks in the Sierra Diablo, on the Diablo Platform, are thin as a result of limited deposition and subsequent erosion.

The Cretaceous section (fig. 9) is dominated by sequences of limestones and sandstones that record transgressive-regressive deposition (Scott and Kidson, 1977). Sandstones were deposited in fluvial-coastal plain (Scott and Kidson, 1977) or beach (Twiss, 1959a) environments, and marine limestones were deposited at various depths of water.

The Yucca Formation, the host of silver-copper-lead deposits in the Indio Mountains, thickens rapidly into the Chihuahua trough. In the Van Horn Mountains, on the flanks of the Diablo Platform, the formation varies in thickness from 0 to 260 ft (80 m) (Twiss, 1959a). Near the prospects in the Indio Mountains, where its base is not exposed, the Yucca Formation is at least 2,070 ft (630 m) thick. In the southern Quitman Mountains, 16 mi (25 km) to the west, closer to the axis of the trough, the Yucca (Mountain) Formation is at least 4,700 ft (1,430 m) thick (Jones and Reaser, 1970) and contains lagoonal, bay, lacustrine, and fluvial strata (Campbell, 1980). The clastic rocks were probably derived from Paleozoic rocks exposed in highlands to the east and southeast.

In this study, the Yucca Formation in the Indio Mountains is informally subdivided into four mappable members: Y1—basal conglomerate (fig. 10a), Y2—conglomerate interbedded with minor amounts of sandstone (fig. 10b), Y3—sandstone interbedded with minor amounts of conglomerate (fig. 10c), and Y4—sandstone and sandy siltstone exhibiting nearshore marine features including clay drapes (fig. 10d), ripple marks (fig. 10e), and burrows (fig. 10f). An 8-ft (2.5-m)-thick limestone marker bed occurs near the top

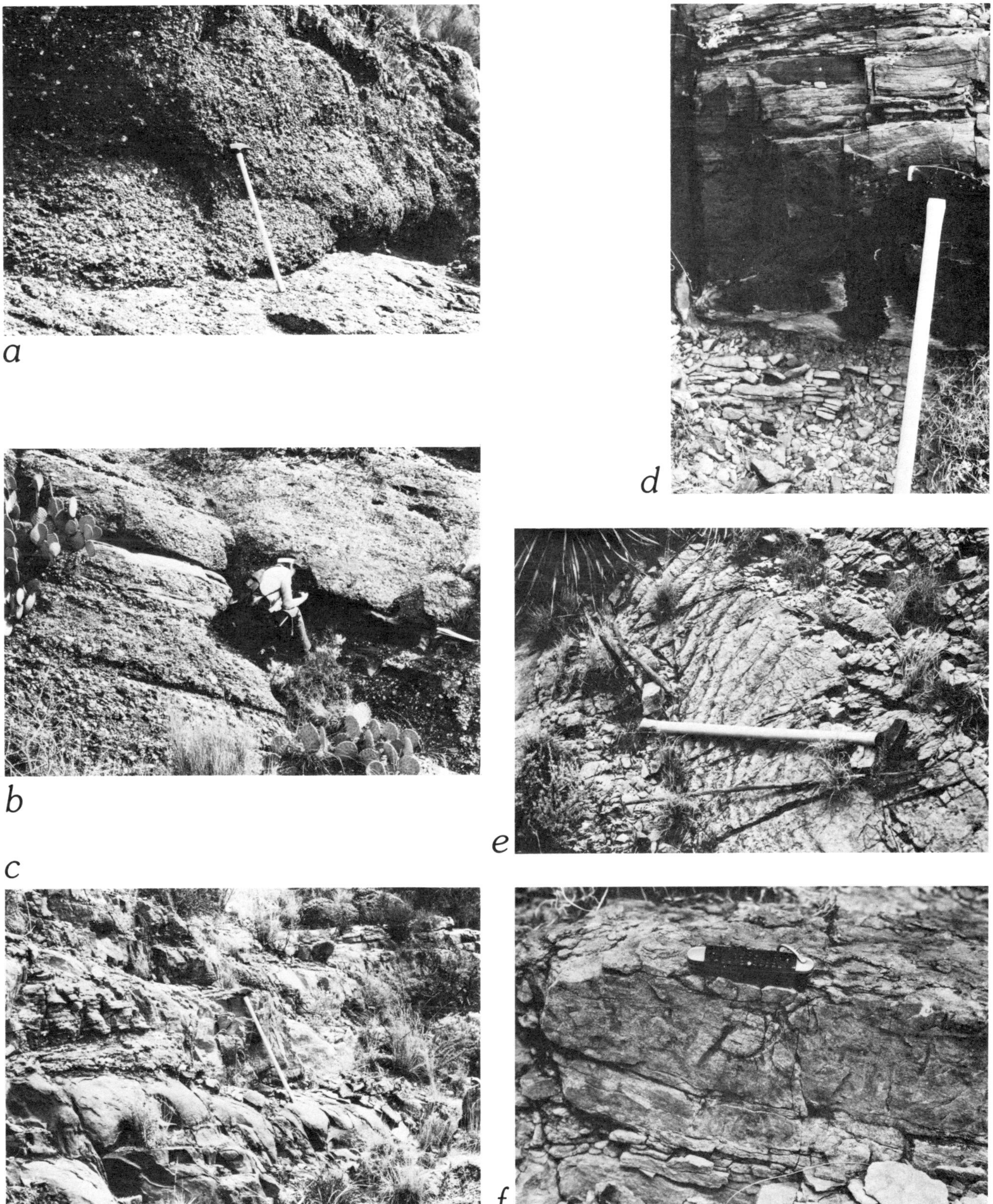

FIGURE 10. Yucca Formation, central Indio Mountains. (a) Red conglomerate of basal member Y1. (b) Red conglomerate and interbedded sandstone of member Y2. (c) Red sandstone of member Y3, which contains minor amounts of conglomerate. (d) Greenish-gray sandy siltstone with clay drapes in upper member Y4. (e) Gray ripple-marked siltstone in Y4. (f) Gray, burrowed sandstone in Y4. For scale, hammer is 2.6 ft (0.8 m) and knife is 3.5 inches (9 cm) long.

of Y4. The overall upward-fining sequence in the Yucca Formation records a transition from fluvial to marine deposition.

Cenozoic Rocks

The Tertiary was a time of major magmatism in Trans-Pecos Texas. Volumetrically significant mafic to silicic volcanic and intrusive activity was restricted to late Eocene and Oligocene time, between 38 and 28 m.y.a. (Henry and McDowell, 1982). Numerous calderas were the dominant volcanic sources (Henry and Price, 1984), producing ash-flow tuffs, lava flows, intrusions, and locally thick sequences of volcaniclastic sediments. Three calderas erupted in the Van Horn area: the Eagle Mountains caldera, the southern boundary of which is 7 mi (11 km) north of the Indio Mountains prospects; the Van Horn Mountains caldera, the northern boundary of which is 3 mi (5 km) southwest of the Plata Verde mine; and a caldera centered in the Wylie Mountains (fig. 1). With the exception of one narrow trachyte dike near the Plata Verde mine (Price, 1982), no Tertiary igneous rocks are known to occur in the immediate vicinity of the silver-copper-lead deposits. This observation suggests that igneous processes were probably not directly involved in ore deposition.

Gravel, sand, silt, and clay fill the bolsons or grabens that began forming in late Oligocene or early Miocene time. In the Van Horn area, thickness of bolson deposits locally exceeds 1,100 ft (335 m). In this part of the Basin and Range province, grabens generally trend north-northwest or north. Deposits of Quaternary alluvium are thin in the Van Horn area.

Limited Lithologic Control of Mineralization

Several similarities exist among the three formations that host the silver-copper-lead deposits. Each formation is composed dominantly of coarse red sandstones or conglomerates, or both, and each fines upward. Carbonate rocks occur within (Hazel) or near the top of the (Powwow and Yucca) red-bed sequences. Carbonate strata commonly underlie and overlie the clastic rocks. However, ore generally does not occur in the carbonate strata. Acidic, carbonate-reactive, hydrothermal fluids of igneous origin are, therefore, unlikely sources for these mineral deposits in the red-bed sequences.

Although the ores are restricted to the red-bed sequences, ores are not confined to individual stratigraphic horizons within the sequences. Hazel-type ores in Precambrian red beds occur along steeply dipping veins that crosscut the nearly horizontal strata (fig. 2). Whereas individual pods of ore at the Plata Verde mine follow bedding in the Permian rocks and whereas geochemical similarities to red-bed copper ores do exist, faults may have controlled epigenetic introduction of sulfides into the formation (Price, 1982). At some of the prospects in Cretaceous rocks, copper, lead, and zinc minerals are dispersed in individual sedimentary layers, but most of the occurrences are along steeply dipping veins. All four mapped subdivisions of the Yucca Formation contain veins (fig. 8).

Structure and Tectonics

Precambrian Deformation

Several Precambrian deformations affected rocks in the Van Horn area (table 4). Two or more periods of deformation produced folds having northeast-trending axes and south-southeast- and southeast-dipping foliation in the Carrizo Mountain Group. Several major tectonic events may have been more or less coincident approximately 1,000 m.y.a.: metamorphism of the Carrizo Mountain Group, northward thrusting of the Carrizo Mountain Group over the Hazel Formation, folding and low-grade metamorphism of the Allamoore and Hazel Formations, and granite intrusion and associated extrusion of rhyolitic ash-flow tuffs northwest of Van Horn (Denison, 1980; Thomann, 1981).

Metamorphic grade in the Carrizo Mountain Group increases from greenschist facies in the north to amphibolite facies in the south. Using K-Ar dates from metamorphic minerals, Denison (1980) estimated the time of metamorphism to be 1,000 ± 25 m.y.a. Thrusting and folding of the Allamoore and Hazel Formations (into dominantly east-trending folds) probably accompanied northward thrusting of the Carrizo Mountain Group along the Streeruwitz Fault (King and Flawn, 1953). Deformation and metamorphism of the Allamoore and Hazel Formations are most intense near the Streeruwitz Fault. Metamorphism was sufficient to produce talc from carbonate rocks in the Allamoore Formation and graphite in the Blackshaft stratigraphic unit in the Hazel Formation. Less than 1 mi (1.6 km) north of the Blackshaft mine (fig. 2), the Hazel Formation is unmetamorphosed and only slightly tilted.

Relatively minor tectonic events occurred near the end of the Precambrian. Left-lateral strike-slip displacement of folded Hazel Formation strata

TABLE 4. Tectonic history of the Van Horn area, Trans-Pecos Texas, modified from King (1965).

		Regional structures parallel to mineralized and unmineralized fractures recognized at silver-copper-lead mines			
Age	**Deformation**	**East-Northeast**	**Northeast**	**Northwest**	**West-Northwest**
Middle Tertiary through Recent	Basin and Range extension; normal faulting, some strike-slip displacement, right-lateral divergent wrenching along west-northwest- and east-striking fractures; two or more orientations of principal stress axes through time	Normal faults	Normal faults	Normal faults	Normal and strike-slip faults
Oligocene	Intrusion and caldera formation; hydrothermal veining resulting from igneous activity	Dikes and hydrothermal veins			
Late Cretaceous through early Tertiary	Laramide folding and thrusting in Chihuahua tectonic belt; some strike-slip faulting - left-lateral convergent wrenching along west-northwest- and east-striking fractures; two orientations of principal stress axes through time	Early strike-slip motion and later tension perpendicular to major fold axes	Early extension perpendicular to major fold axes and later strike-slip motion	Major fold axes in Indio and Van Horn Mountains	Major fold axes in Sierra Diablo, strike-slip faults
Permian through middle Mesozoic	Monoclinal tilting along west-northwest-striking flexures				Flexures
Late Pennsylvanian through Early Permian	Ouachita-Ancestral Rockies folding and thrusting in Marathon region to southeast; uplift and erosion near Van Horn; dip-slip displacement along steep west-northwest-striking faults in Sierra Diablo	Trend of folds in Marathon region		Possible extension perpendicular to trend of folds	Dip-slip faults
Between latest Precambrian and Ordovician	Pre-Bliss Sandstone, post-Van Horn Sandstone tilting and local faulting		Dip-slip fault at Dallas prospect		
Later Precambrian	Left-lateral strike-slip displacement on the Grapevine Fault				Strike-slip faults
Late Precambrian	Folding and faulting of the Allamoore and Hazel Formations; northward thrusting of the Streeruwitz thrust sheet	Fold axes near Blackshaft mine		Extension perpendicular to folds	
Precambrian	Two or more periods of deformation of the Carrizo Mountain Group	South-southeastward-dipping foliation	Southeastward-dipping foliation		

took place along a west-northwest-striking fault (King and Flawn, 1953; King, 1965). Faults having this orientation, which constitute the Texas lineament trend defined by Muehlberger (1980), moved repeatedly throughout Phanerozoic time (table 4). Latest Precambrian(?) uplift and erosion led to deposition of the Van Horn Sandstone on a highly dissected landscape. Dip-slip motion on the fault along which the Dallas prospect is located (fig. 2) occurred between Late Precambrian and Ordovician time (King, 1965). Erosion during this time leveled the region to a peneplain on which Ordovician rocks were deposited (Denison, 1980).

Paleozoic Deformation

Several unconformities having little or no angular discordance attest to mild, probably epeirogenic movements during the early Paleozoic (Harbour, 1972; Muehlberger, 1980). Variations in thickness of Cambrian through Devonian rocks (Galley, 1958; LeMone and others, 1983) indicate that both the Tobosa Basin, the forerunner of the Delaware Basin, and the Diablo Platform, which became prominent during late Paleozoic time, had early Paleozoic precursors.

Major Paleozoic deformation was associated with the Ouachita - Marathon fold and thrust belt, which is exposed in the Marathon and Solitario areas 110 mi (170 km) southeast of the Van Horn area. Deformation probably began with uplift in Mississippian time (King, 1937; Galley, 1958) and culminated in thrusting, folding, and uplift in Late Pennsylvanian - Early Permian time. Thrusting was from the southeast to the northwest (King, 1937); the youngest rocks involved are Early Permian (early Wolfcampian) (King, 1980).

In the Van Horn area, foreland deformation contemporaneous with the Ouachita - Marathon thrust belt yielded generally north-trending arches and troughs (King, 1965), including the Diablo Platform and the Delaware Basin. West-northwest-striking faults in the Sierra Diablo underwent down-to-the-north, dip-slip displacement at that time. The relationship between foreland deformation, which included uplifts of the ancestral Rocky Mountains, and contemporaneous plate collisions is speculative (Goldstein, 1981; Kluth and Coney, 1981). Wolfcamp-age Powwow conglomerates, which rest unconformably on various older rocks, attest to uplift and erosion that accompanied the Ouachita - Marathon deformation.

In the Sierra Diablo north of the Van Horn area, post-Wolfcampian Permian deformation produced down-to-the-north, north-northwest-trending flexures, which probably controlled the location of limestone reefs during Leonardian time (King, 1965). Tilting of Permian rocks occurred before deposition of discordant Lower Cretaceous sedimentary rocks (King, 1980).

Mesozoic - Cenozoic Activity

Laramide Deformation

Laramide deformation involved sliding of Cretaceous rocks in the Chihuahua trough along a décollement zone in underlying Jurassic evaporites and thrust faulting, folding, and monoclinal warping along the eastern margin of the trough. The stable Diablo Platform acted as a buttress so that indicators of the most pronounced structural activity, such as low-angle thrusts and overturned folds, are generally located along the boundary between the platform and the trough.

The major episode of Laramide folding occurred in late Paleocene time in the Big Bend area (Wilson, 1971) and probably affected all of Trans-Pecos Texas. Wilson (1971) showed that sedimentologic evidence of deformation first appeared in lower Eocene deposits. Regional uplift without folding probably occurred, and erosion surfaces developed at the end of the Early Cretaceous and at several times during the Late Cretaceous (Maxwell and others, 1967). Laramide deformation in general is believed to have resulted from Farallon - North American plate convergence (Dickinson, 1981). Low-angle subduction of the Farallon plate during the Late Cretaceous and early Tertiary may have led to shear along the base of the overlying North American plate. The significance of this shear to Laramide deformation is unknown.

During Laramide deformation in Texas and elsewhere, the maximum principal compressive stress, σ_1, was directed east to northeast. In Arizona and New Mexico, Laramide σ_1 was dominantly east-northeast (Heidrick and Titley, 1982). In the Colorado Plateau - Rocky Mountain area of Colorado and New Mexico, early east-northeast compression may have been followed by later northeast compression (Chapin and Cather, 1981). Studies in Trans-Pecos Texas have also revealed two directions of compression: east to east-northeast (Muehlberger, 1980; Moustafa, 1983) and northeast (Berge, 1981; Decamp, 1981), but in Texas northeast compression predates east to east-northeast compression. In the Malone Mountains 40 mi (60 km) west of the Van Horn area, north-trending folds refold northwest-trending folds (Berge, 1981). At Mesa de Anguila 80 mi (130 km) southeast of the Van Horn area, orientations of faults, fractures, folds, stylolites, and calcite twin-lamellae suggest that σ_1 during early Laramide deformation was N50° to 55°E but changed to N65° to 80°E (Decamp, 1981).

Geologic maps and structural measurements made during this study can be interpreted to indicate two periods of extensive Laramide deformation in the Indio Mountains: Laramide folding and thrust faulting and Basin and Range normal faulting (fig. 8). East of the Basin and Range Indio Fault, structures are relatively simple (fig. 11). West of the Indio Fault, outcropping rocks are part of a thrust sheet that has been eroded east of the fault. Thrusting is interpreted to have predated folding (fig. 12).

Measured slickenlines on minor faults (table A1 in the appendix), which are well preserved in the Yucca Formation, have been grouped into four

FIGURE 11. Eastward-dipping Cretaceous sedimentary rocks east of the Indio Fault, Indio Mountains. Prominent ledges in the area shown in the middle of the photograph are resistant limestones of the Bluff Formation, which overlies the Yucca Formation and is underlain by the Cox Sandstone.

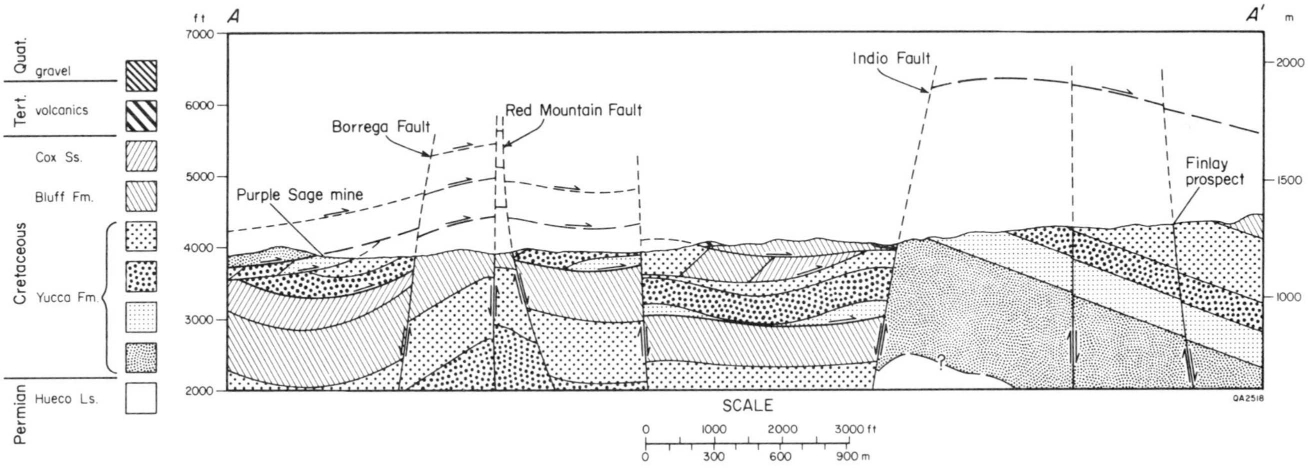

FIGURE 12. Cross section A-A', illustrating major structural features in the Indio Mountains such as Laramide thrusts and folds cut by Basin and Range normal faults. Note that outcropping rocks in the horst between the Borrega and Red Mountain normal faults are younger than outcropping rocks on either side of the horst. Laramide thrusting placed older rocks on top of younger rocks, Basin and Range extension produced the horst, and erosion removed the upper plate rocks in the horst. See figure 8 for location of section.

probable phases of deformation (table 5). Approximately half of the slickenline data listed in table A1 could correspond to more than one of the probable deformations and were not analyzed further. Data from table 5, which are plotted in figure 13, indicate two Laramide deformations having σ_1 oriented northeast and east-northeast. Relative timing of these two events was not established in this study but was assumed to be the same as that deduced by Berge (1981) and Decamp (1981). Scatter in the data shown in figure 13 can be interpreted as the result of a gradual clockwise rotation in σ_1 during Laramide deformation. S. M. Cather (personal communication, 1983) suggested that σ_1 may actually have been oriented east-northeast throughout the deformation in the Indio Mountains and that the difference in σ_1 between early and late Laramide compression is due to large-scale counterclockwise rotation resulting from left-lateral strike-slip movement along west-northwest basement faults of the Texas lineament (Muehlberger, 1980). More detailed structural

TABLE 5. Slickenline data that correspond to only one of four probable phases of deformation, Indio Mountains.[1]

Strike of fault	Type of motion[2] and (number of faults measured): Laramide compression		Basin and Range tension	
	Early σ_1 NE	**Late** σ_1 ENE	**Early** σ_3 ENE	**Late** σ_3 NW
N	reverse, right-lateral strike-slip (2)			
NNE				normal, right-lateral dip-slip (4)
NE		reverse and normal, right-lateral strike-slip (4)	normal, left-lateral dip-slip (2)	normal, right-lateral dip-slip (1)
ENE	reverse and normal, left-lateral strike-slip (10)			normal, left-lateral dip-slip (2)
E			normal, right-lateral dip-slip (3)	normal, left-lateral dip-slip (2)
WNW		normal and pure, left-lateral strike-slip (3)		normal, left-lateral dip-slip (1)
NW		reverse and pure, left-lateral strike-slip (4)	normal, right-lateral dip-slip (1)	
NNW	reverse, left-lateral dip-slip (1)			normal, right-lateral oblique-slip (1)

[1]Refer to table B1 in appendix for a listing of all slickenline data.
[2]Strike-slip fault is defined as having a rake or slip-line pitch less than or equal to 30°; dip-slip fault is defined as having a rake greater than or equal to 60°.

studies in the Indio Mountains and elsewhere in the Trans-Pecos region may clarify the orientation of Laramide stresses at different times.

Middle Tertiary Magmatism

Price and Henry (1984) demonstrated that magmatism before about 30 m.y.a., including most caldera-forming volcanism, occurred while the area was under east-northeast- to east-oriented compression remaining from Laramide deformations. The Trans-Pecos volcanic rocks, however, were not involved in Laramide folding and thrusting. Regionally persistent orientations of veins and dikes (fig. 14) indicate that σ_3 was probably directed north-northwest to north between 41 and 32 m.y.a. (unless otherwise noted, ages are based on K-Ar dating, chiefly of mineral separates, summarized by McDowell [1979] and Price and Henry [1984]). The igneous rocks were formed in a continental arc related to subduction along a paleotrench that lay off the coast of western Mexico. A transition from compression to tension occurred about 30 m.y.a., and subsequent volcanism, including the San Carlos - Santana caldera complex in Chihuahua (Henry and Price, 1984) and minor basalts in Texas, is related to Basin and Range extension.

Basin and Range Extension

Extension in the Texas part of the Basin and Range province began about 30 m.y.a., but normal faulting probably began several million years

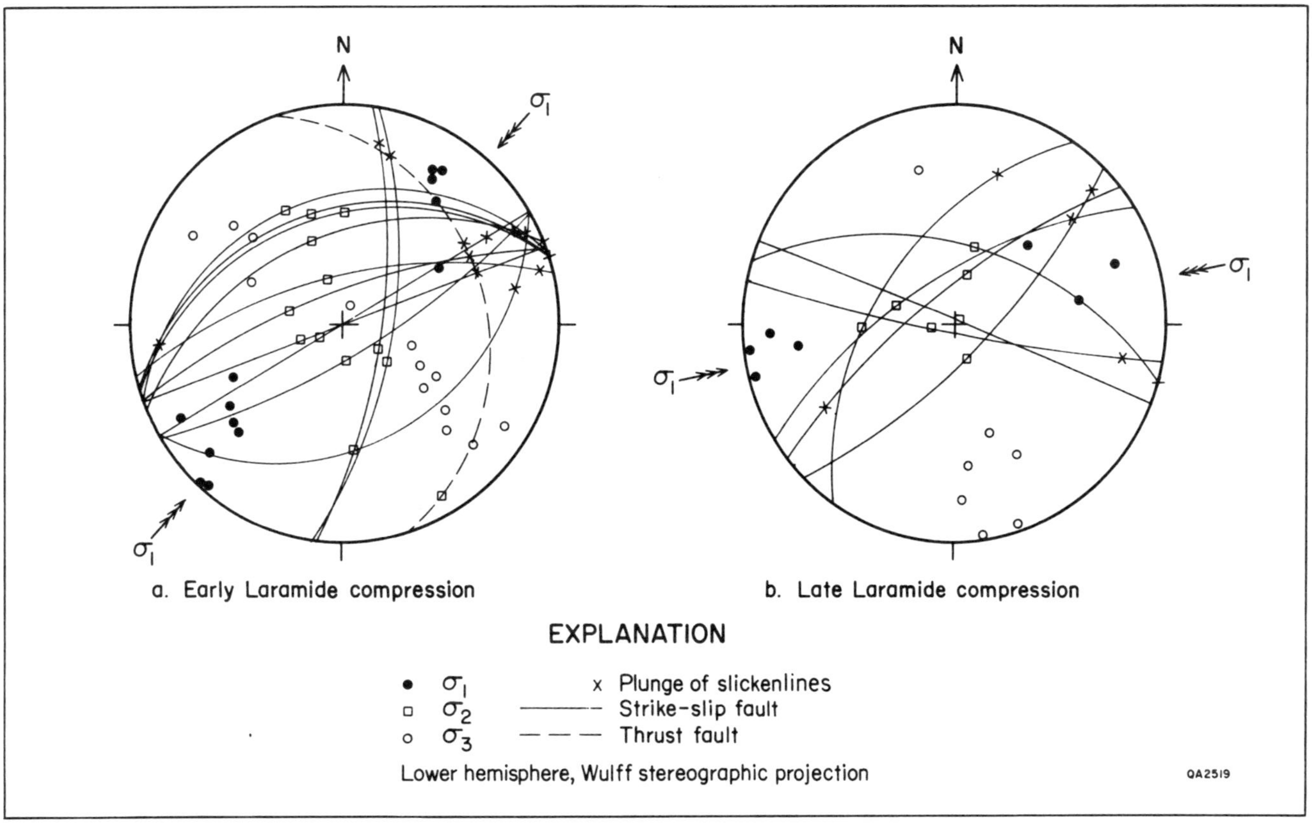

FIGURE 13. Orientations of faults, slickenlines, and corresponding principal stress axes, Laramide deformation, Indio Mountains. The intermediate principal compressive stress, σ_2, is assumed to be in the fault plane and perpendicular to the slickenlines, which form a 30° angle to the maximum principal compressive stress, σ_1. (a) Northeast-directed early Laramide compression occurred along a more or less horizontal σ_1, and resulted in right-lateral strike-slip motion on north-striking shear fractures and left-lateral strike-slip motion on east-northeast-striking shear fractures. (b) East-northeast-directed late Laramide compression occurred along a nearly horizontal σ_1, and resulted in right-lateral strike-slip motion on northeast-striking fractures and left-lateral strike-slip motion on west-northwest-striking fractures. East-striking faults having left-lateral strike-slip displacement are not shown because they could have formed during either the early or late Laramide compression or during an intermediate time. Northwest-striking faults having left-lateral strike-slip motion that occurred during late Laramide compression are not shown because these faults were probably reactivated early Laramide structures for which the assumption that σ_2 is in the fault plane is invalid.

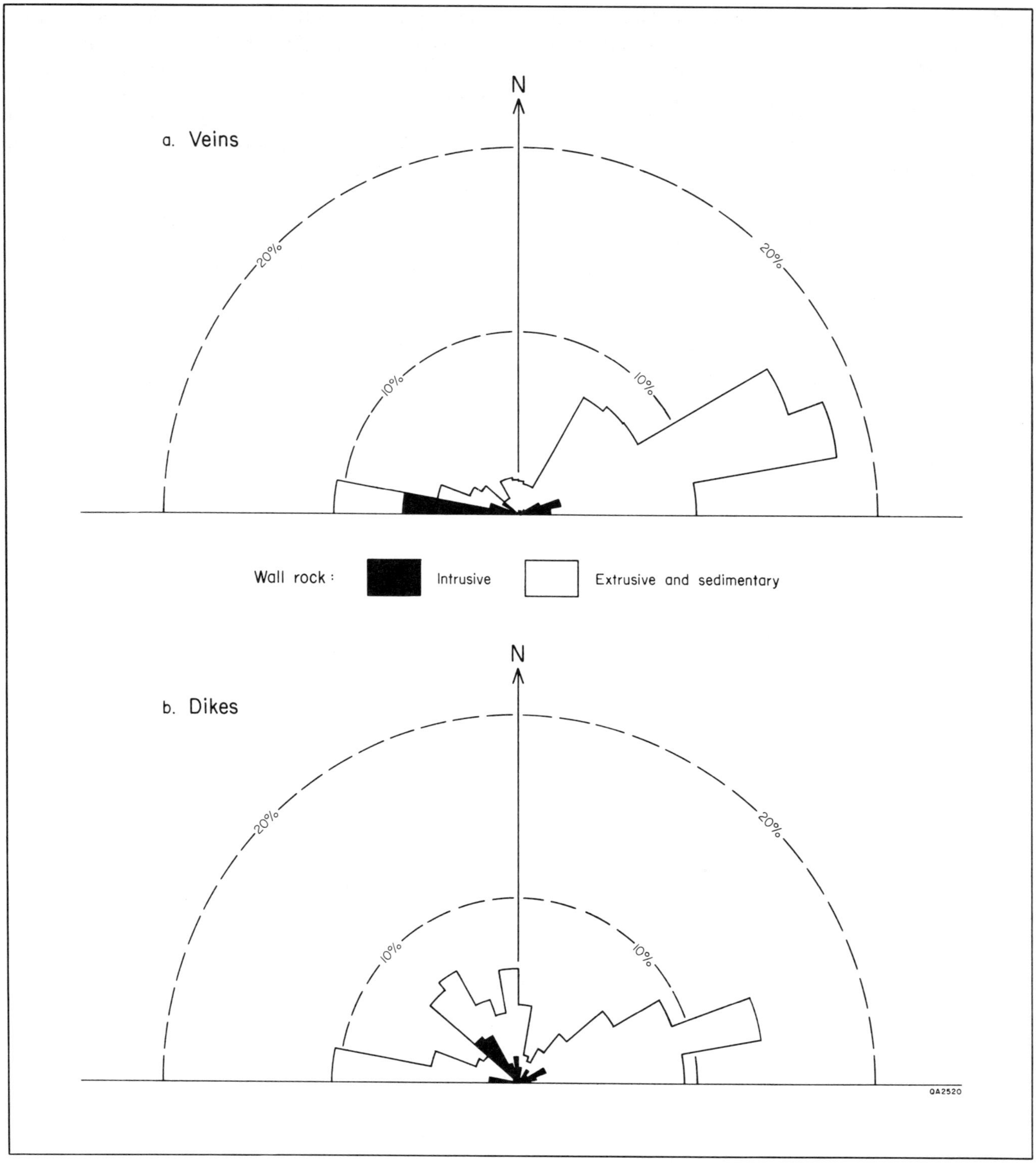

FIGURE 14. Strike rosettes of (a) veins and (b) dikes that were emplaced during the major period of igneous activity, 32 to 41 m.y.a., in the Trans-Pecos region. Strike bars are in 10° intervals. Lengths of bars indicate percentages of the total strike length: (a) 46 mi (74 km), representing 866 veins in the Eagle, Quitman, Organ, and Chinati Mountains and the Shafter silver and Terlingua mercury districts, and (b) 109 mi (175 km), representing 394 dikes in the Eagle, Indio, Quitman, Organ, and Chinati Mountains, Paisano volcano, Pine Canyon and Infiernito calderas, and Big Bend National Park. The dominant east-northeast to east strikes indicate that the least principal stress axis, σ_3, during this time was directed north-northwest to north. Data are compiled from mapping by Dunham (1935), Ross (1943), Gillerman (1953), Yates and Thompson (1959), Underwood (1963), Albritton and Smith (1965), Maxwell and others (1967), Laux (1969), Jones and Reaser (1970), McAnulty (1972), Parker (1976, 1979), Ogley (1978), Murry (1979), Laughlin and others (1982), Cepeda and Henry (1983), and Henry and Price (in progress).

later. Regional east-northeast to east dike and vein orientations indicate that the least principal compressive stress, σ_3, was directed north-northwest to north until about 32 m.y.a. (fig. 14). Shortly thereafter, regional dike orientations changed dramatically. Approximately 100 mi (160 km) southeast of the Van Horn area, a generally north-striking, 32-m.y.-old, 4-mi-(7-km)-long basaltic dike cuts Cretaceous rocks near the 30- to 28-m.y.-old San Carlos - Santana caldera complex (Chuchla, 1981; Gregory, 1981). North-northwest-striking dikes occur in and near the 28- to 26-m.y.-old, dominantly basaltic Bofecillos volcano (McKnight, 1970; McDowell, 1979). This change in dike orientation records a change from east-northeast compression to east-northeast-oriented Basin and Range tension. In New Mexico, the estimated age of earliest extension is also 30 m.y. (Chapin, 1979; Zoback and others, 1981; Laughlin and others, 1983).

The 30- to 26-m.y.-old volcanic rocks in Texas appear to have been erupted before Basin and Range faulting began. By 23 to 20 m.y.a., when the Rim Rock dikes were intruded (Dasch and others [1969]; recalculated by Laughlin and others [1982]) immediately south of the Van Horn area, faulting had begun. Some Rim Rock dikes formed along major normal faults and fed flows that were interlayered with bolson fill (Dasch, 1959).

Dasch and others (1969) estimated the ages of Rim Rock dikes corresponding to two dominant strike orientations. Assuming that the dikes were injected perpendicular to σ_3, Dasch and others argued that west-northwest-trending dikes were injected about 23 m.y.a. when σ_3 was north-northwest and that north-northwest-trending dikes were injected about 20 m.y.a. when σ_3 was oriented east-northeast. Although field relations indicate that some north-northwest dikes are younger than some west-northwest dikes, the samples of west-northwest dikes chosen for whole-rock K-Ar dating by Dasch and others (1969) were too altered to prove that the two trends are significantly different in age. Both sets of dikes are petrographically similar alkali basalts locally containing xenocrysts of alkali feldspar, plagioclase, biotite, amphibole, pyroxene, apatite, and magnetite; they were probably derived from the same magma source.

An alternative explanation of Rim Rock dike trends is that the two trends are more or less contemporaneous and that the volumetrically dominant north-northwest set (fig. 15) was injected perpendicular to an east-northeast-

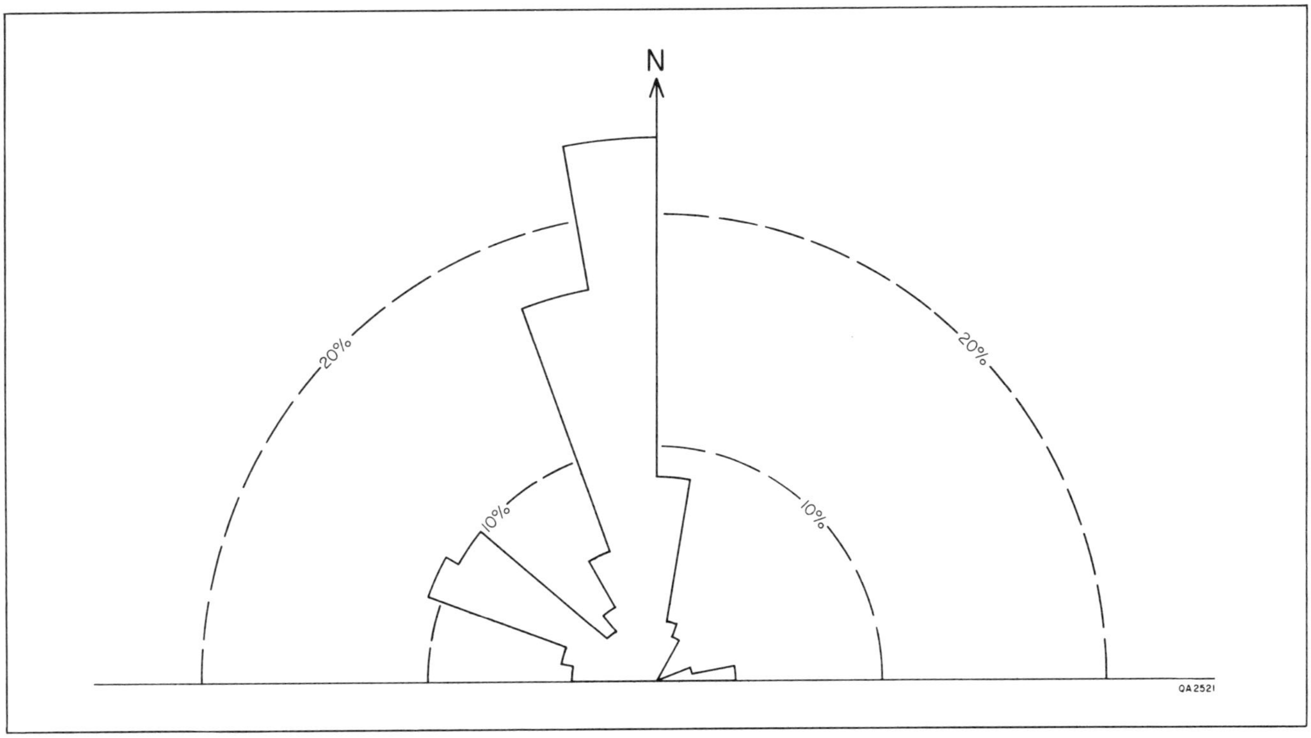

FIGURE 15. Strike rosettes of early Miocene Rim Rock dikes. Strike bars are constructed in 10° intervals. Lengths of bars are proportional to percentages of the 44-mi (70-km) total strike length representing 480 dike segments. Data are compiled from Dasch (1959) and from analysis of aerial photographs.

oriented σ_3 that prevailed about 20 m.y.a. The subsidiary west-northwest dikes may have developed when preexisting basement structures of that trend (Muehlberger, 1980) opened in response to east-northeast tension. This favored explanation is consistent with observations from the San Carlos - Santana caldera complex and the Bofecillos volcano (see page 21) and with the results of other studies conducted throughout the Basin and Range province.

Rehrig and Heidrick (1976) indicated that σ_3 was east-northeast during late Tertiary intrusive activity in Arizona. Zoback and Thompson (1978) concluded that σ_3 was N68°E during injection of a 17- to 14-m.y.-old dike swarm in northern Nevada. Lipman (1981) suggested that σ_3 was oriented northeast during formation of a 23-m.y.-old batholith and associated northwest-trending dikes in the Latir - Questa area of New Mexico. Golombek (1982) concluded that extension in the Española Basin of New Mexico was east-northeast before 10 m.y.a.

Numerous geologists have observed a shift in the Basin and Range province from early east-northeast to later west-northwest extension (Zoback and Thompson, 1978; Lipman, 1981; Zoback and others, 1981; Golombek, 1983; Lucchita and Suneson, 1983; Morgan and Seager, 1983). In New Mexico and perhaps throughout the province, the shift to west-northwest extension occurred about 10 m.y.a. (Golombek, 1982). Although such a shift thus far has not been documented in Texas, it may be recorded by the silver-copper-lead veins in red-bed sequences and by Basin and Range faults in the Indio Mountains. Minor faults in the Indio Mountains are interpreted to have formed not only during two periods of Laramide compression (fig. 13) but also during two periods of Basin and Range extension (table 5 and fig. 16). Relative timing was not established in this study; on the basis of interpretation of regional dike data discussed previously, it was assumed that σ_3 during early Basin and Range extension was directed east-northeast. Late Basin and Range extension appears to have been directed northwest (fig. 16).

Extension in the Basin and Range province is generally attributed to change from a convergent (Laramide) to a transform boundary along the

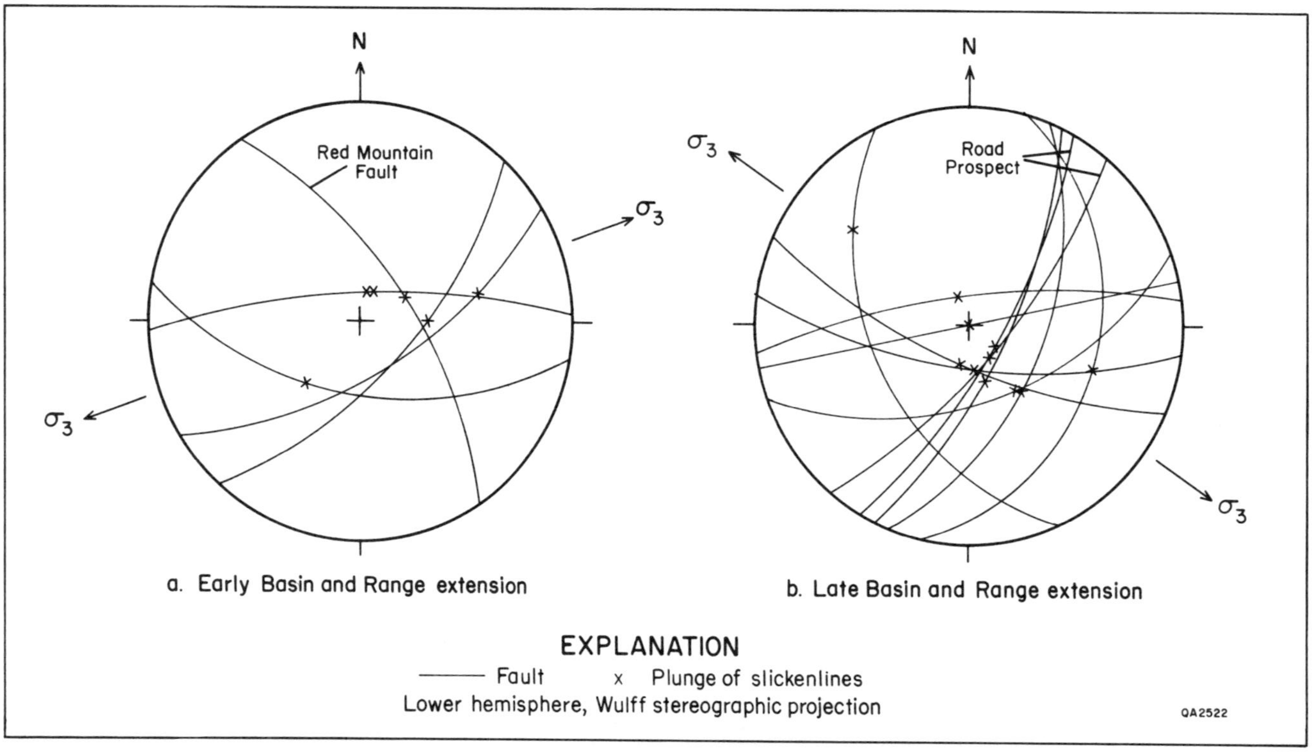

FIGURE 16. Orientations of faults and slickenlines, Basin and Range deformation, Indio Mountains. It is assumed that slickenlines along either new or preexisting fractures occur in the σ_1, σ_3 plane and that σ_3 is horizontal. (a) Early Basin and Range extension occurred in an east-northeast direction, approximately parallel to the direction of late Laramide compression. (b) Late Basin and Range extension occurred in a northwest direction and allowed east-northeast- and northeast-striking fractures in red beds to be opened and filled with silver, copper, and lead minerals.

western edge of North America (Dickinson, 1981; Zoback and others, 1981). Clockwise rotation of σ_3 with time may be related to movement of the transform boundary (Ingersoll, 1982). The Basin and Range province is actively extending today. Quaternary fault scarps are abundant in the Texas part of the province (Muehlberger and others, 1978; Seager, 1980; Henry and others, 1983). In the Van Horn area, active faults occur along the eastern margins of the Van Horn, Carrizo, and Baylor Mountains and in the Sierra Diablo. Many Quaternary faults presumably are reactivated earlier Basin and Range faults; therefore, strikes of scarps cannot be used to uniquely define σ_3.

First-motion studies of the 1931 Valentine earthquake suggest a further change in σ_3 to about N74°E at present (Dumas and others, 1980). This event involved right-lateral strike slip along a west-northwest-striking fault. Another seismic study by Dumas (1981) indicated normal faulting related to a swarm of earthquakes located in Mexico near the Indio Mountains approximately 30 mi (50 km) south-southwest of Van Horn. For these events, σ_3 was N35°W, an orientation more consistent with that typical of late Basin and Range extension elsewhere in the province. More seismic data from the Trans-Pecos region should be collected and analyzed to determine current stress orientations.

STRUCTURAL CONTROLS OF VEIN MINERALIZATION

East-Northeast and Northeast Veins in Precambrian Rocks

Steeply dipping veins yielded most of the silver from Precambrian rocks in the Van Horn - Allamoore district (tables 2 and 3). Von Streeruwitz (1892b) indicated that individual sulfide veins at the Hazel mine follow a vertical fracture zone and diverge upward (fig. 17). Veins in the district display two dominant strike orientations: east-northeast and northeast (fig. 2). Contemporaneous filling of fractures with veins of these two trends is suggested by the close spatial associations and the identical mineralogical compositions of the two trends. At the Mohawk mine, fractures of both trends are abundant, and along the length of the surface exposure (fig. 18) silver-copper mineralization switches back and forth from the northeast to the east-northeast trend. On a smaller scale, evidence of contemporaneous opening of east-northeast and northeast fractures is visible at the Hazel mine (fig. 19). Also at the Hazel mine, wider sulfide mineralization occurred at the intersection of east-northeast and northeast fractures (Flawn, 1952).

Four major trends of nearly vertical fractures were recognized in the Hazel Formation during this study: east-northeast and northeast ore trends plus northwest- and west-northwest-striking fractures. That fractures of the ore trends were opened more than those of the other trends is shown not only by the paucity of ore in the northwest and west-northwest trends but also by the dominance of calcite fillings (unrelated to ore) along the east-northeast and northeast trends (table 6).

Relative timing of fractures of different trends was determined at several localities near the Hazel-type veins. Younger fractures hook or curve into and terminate against older fractures (fig. 20). East-northeast and northeast trends are, in general, younger than the northwest and west-northwest trends. Geologic relations preclude unambiguous determination of age of fracturing and mineralization relative to major stratigraphic or tectonic events. Regional geologic maps by King (1965) demonstrate that all four fracture trends were active during Basin and Range extension (table 4). Although mineralization in the

TABLE 6. Percentages of fractures of the four major trends in the Hazel Formation that exhibit extensional features, ore-related filling and calcite filling not associated with mineralization.

Percentage of fractures with fillings	Fracture trend and number of fractures measured			
	ENE (100)	NE (107)	NW (60)	WNW (50)
Ore-related copper oxides, sulfides, barite, or limonite	19	19	5	2
Calcite	21	14	3	6
Total	40	33	8	8

Precambrian rocks could have occurred earlier, the following discussion presents a case for ore deposition during Basin and Range extension.

To determine the age of mineralization, fractures in Permian rocks overlying the Hazel Formation were examined as close to the veins as possible (fig. 6). If the Hueco Limestone exhibits the same trends of fractures as the Hazel Formation, then the fractures in both host rocks are probably younger than the Permian rocks. Measurements in the field indicate that this is, in fact, the case for the east-northeast and northeast ore trends (fig. 21), the trends determined to be younger on the basis of fracture intersections. Silver-copper-lead mineralization therefore most likely occurred during one of the Tertiary post-Hueco deformations (table 4): Laramide shortening, middle Tertiary magmatism, or Basin and Range extension. W. R. Muehlberger (personal communication, 1984) suggested that the east-northeast-trending fractures at the Mohawk and Hazel mines (fig. 2) formed as a set of left-lateral, en echelon tension fractures during Laramide deformation. Although the fractures may have formed at this time, mineralization could have occurred much later.

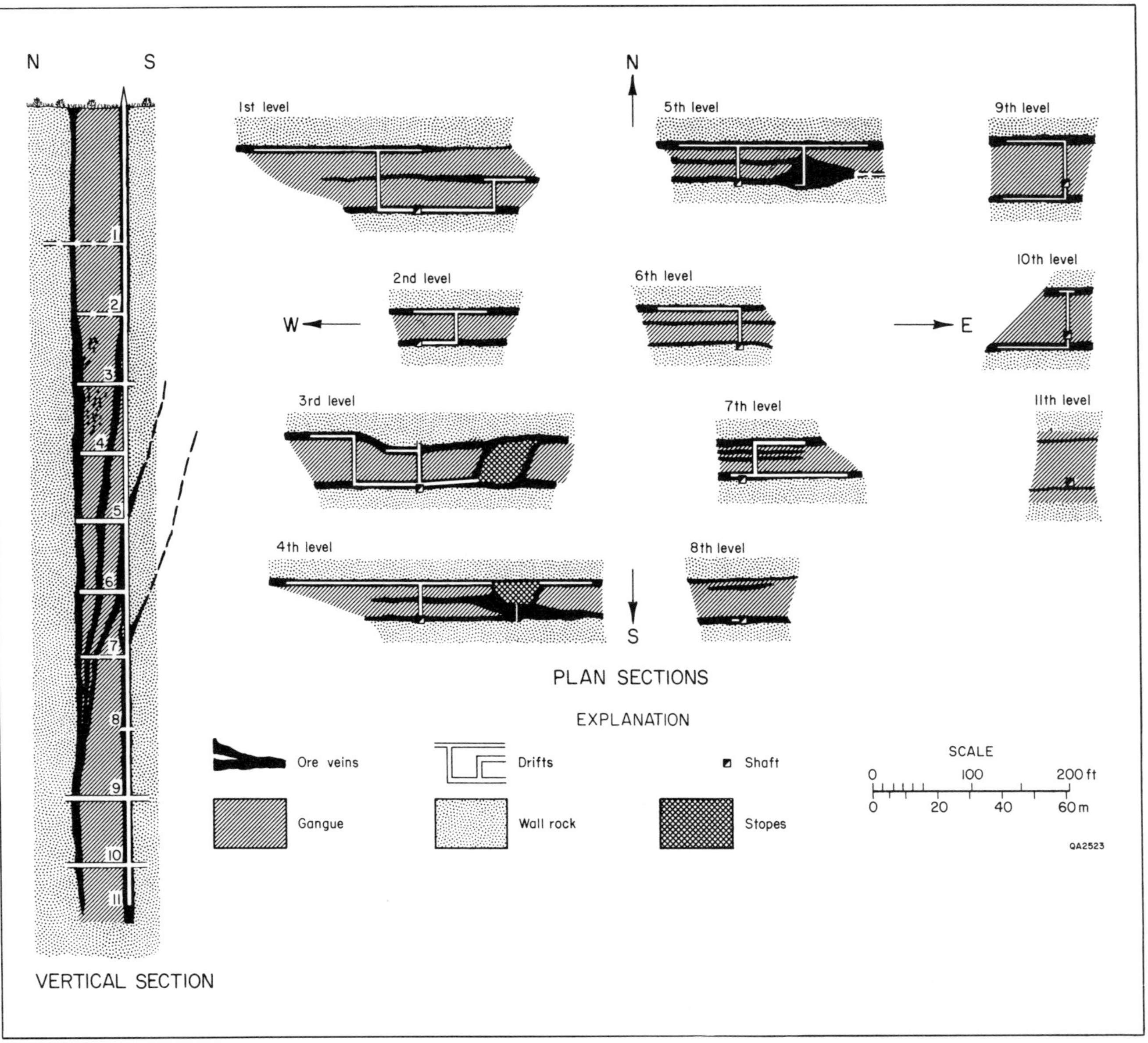

FIGURE 17. Veins at the Hazel mine in 1891 sketched by von Streeruwitz (1892b). Gangue mapped by von Streeruwitz is brecciated, gray material from the Hazel Formation containing minor amounts of sulfides and local veinlets of barite and calcite.

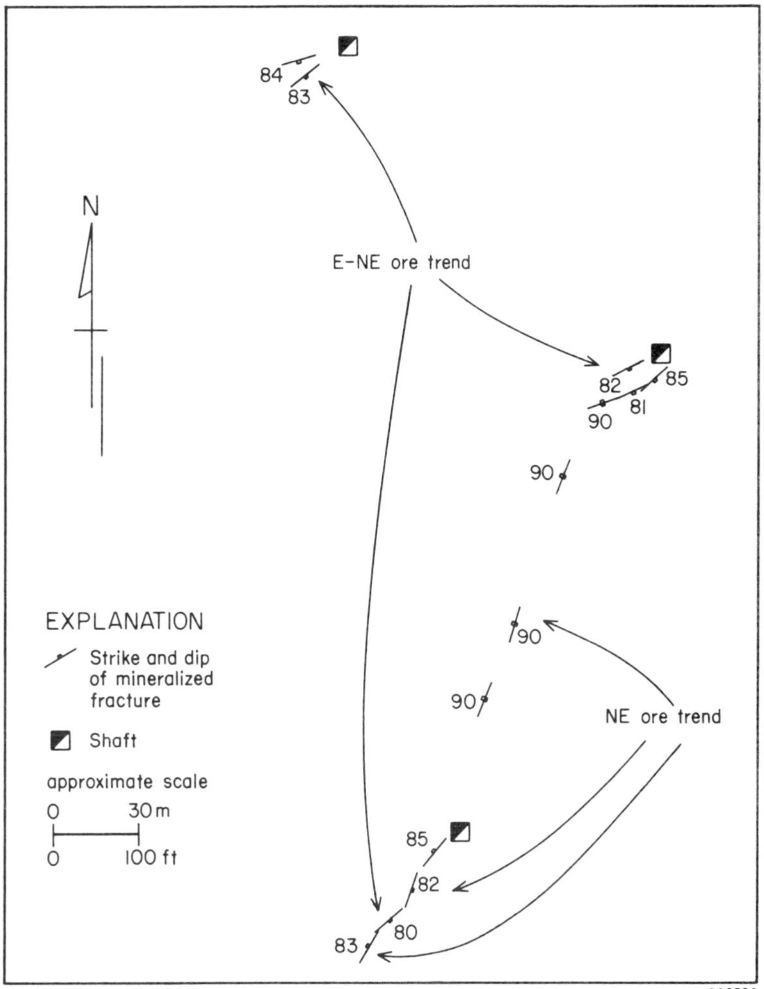

FIGURE 18. Orientations of mineralized fractures at the Mohawk mine. In map view, veins switch back and forth from the northeast to the east-northeast ore trends, suggesting contemporaneous opening of both sets of fractures. Veins of both trends are made of the same material: barite, calcite, and oxidized copper minerals.

FIGURE 19. Calcite-filled fractures in a streambed at the Hazel mine. Acid bottle is 3 inches (77 mm) long. Wider calcite veins are along fractures of the east-northeast ore trend. Calcite filling connects one vein to another along a fracture of the northeast ore trend, which suggests that fractures of both trends were open at the same time.

FIGURE 20. Typical fracture intersections in the Hazel Formation. Fractures trending from right to left terminate against and are younger than the long fracture trending from lower right to upper left.

FIGURE 21. (right) Orientations of fractures at mines in the Hazel Formation and at nearby exposures in the Hueco Formation. (a) Poles to 89 mineralized and unmineralized fractures (lower hemisphere, Schmidt equal-area projection) in the Hazel Formation at the Marvin - Judson prospect. Limonite staining occurs along fractures of the Hazel ore trend at this prospect. (b) Poles to 67 unmineralized fractures in the Hueco Formation 2,600 ft (790 m) north-northwest of the Marvin - Judson prospect. (c) Poles to 117 mineralized and unmineralized fractures in the Hazel Formation at the Hazel mine. Silver-copper mineralization occurs dominantly along east-northeast-striking fractures and to a lesser extent along northeast-striking fractures. (d) Poles to 56 unmineralized fractures in the Hueco Formation 2,200 ft (670 m) north of the Hazel mine. (e) Poles to 76 mineralized and unmineralized fractures in the Hazel Formation at the Pecos mine, and Diablo and Eureka prospects. Silver-copper-lead mineralization occurs exclusively along northeast-striking fractures. (f) Poles to 55 unmineralized fractures in the Hueco Formation 2,300 ft (700 m) west-northwest of the Pecos mine.

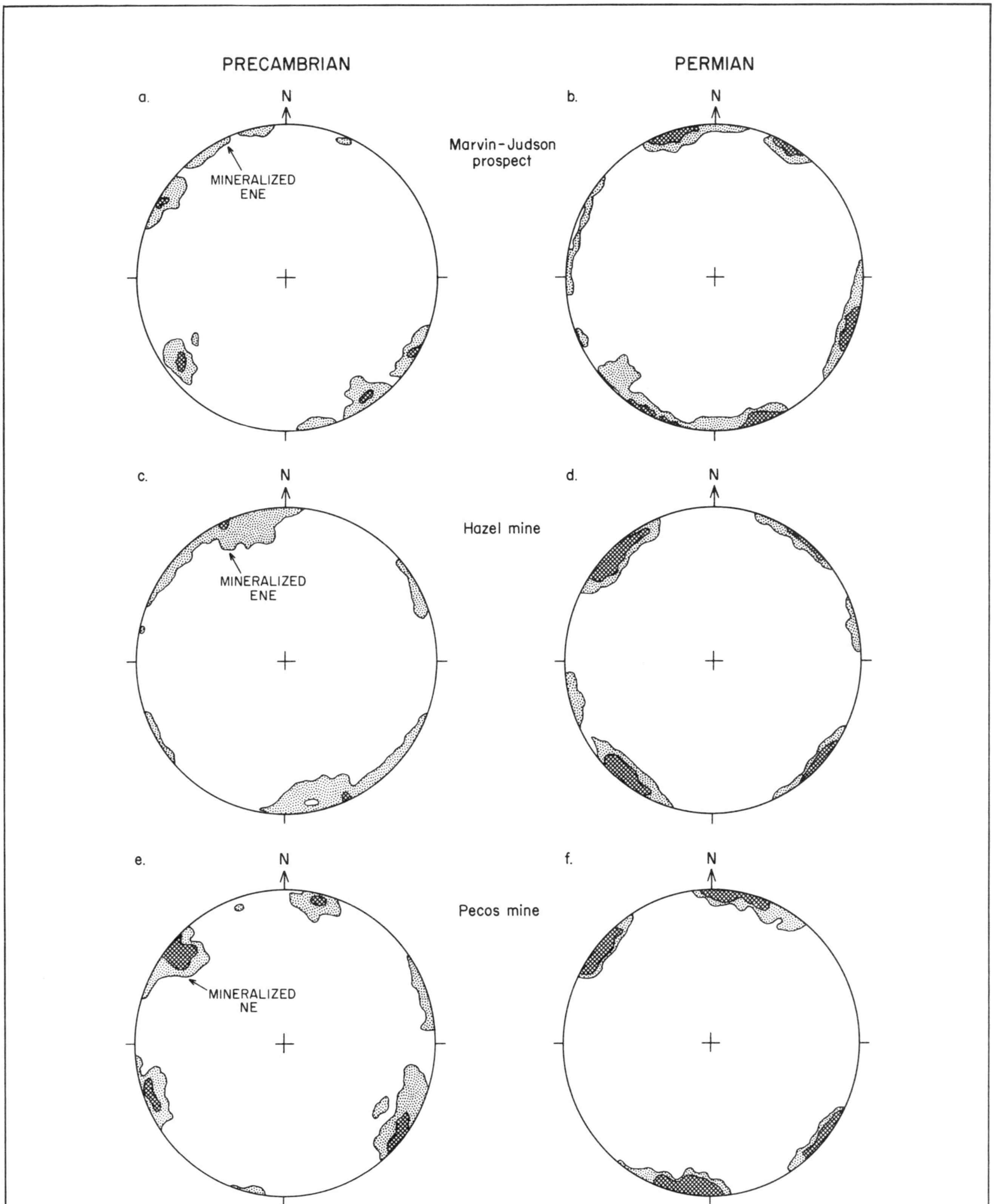
PRECAMBRIAN
PERMIAN
a.
N
MINERALIZED ENE
b.
N
Marvin-Judson prospect
c.
N
MINERALIZED ENE
d.
N
Hazel mine
e.
N
MINERALIZED NE
f.
N
Pecos mine
4-10%
>10% per 1% area
HAZEL FM.
HUECO LS.
QA2525

East-Northeast and Northeast Veins in Cretaceous Rocks

Veins in the Yucca Formation of the Indio Mountains have essentially the same orientations as those in the Hazel Formation of the Van Horn - Allamoore district (fig. 22a and c). Although Laramide and Basin and Range deformations in the Indio Mountains produced numerous fractures (figs. 8 and 22d), only those having east-northeast or northeast strikes are mineralized. Clearly some east-northeast- and northeast-striking shear fractures developed during Laramide compression (table 5). Displacements of beds and slickenline directions indicate that faults at the Road and Finlay prospects (fig. 8) are normal and probably moved during late Basin and Range extension (fig. 16), however.

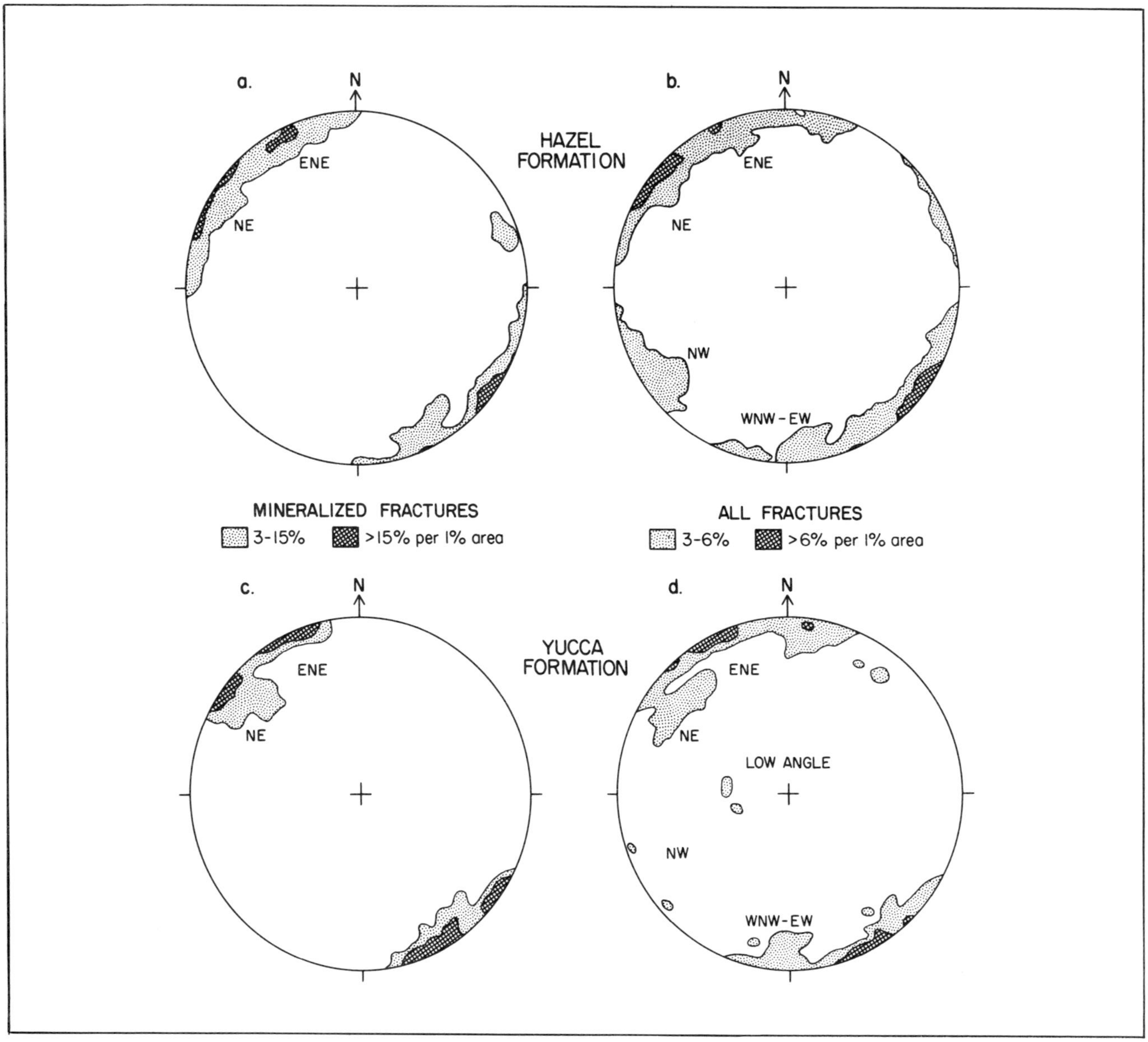

FIGURE 22. Fracture patterns at and near silver-copper-lead deposits in the Precambrian Hazel Formation and in the Cretaceous Yucca Formation. Mineralization in both formations occurs chiefly along high-angle fractures having east-northeast and northeast strikes. Abundant high-angle fractures having northwest and west-northwest to east strikes and low-angle fractures are generally unmineralized. (a) Poles to 46 mineralized fractures in the Hazel Formation, southern Sierra Diablo area (lower hemisphere, Schmidt equal-area projection). (b) Poles to 342 mineralized and unmineralized fractures in the Hazel Formation. (c) Poles to 33 mineralized fractures in the Yucca Formation, Indio Mountains. (d) Poles to 167 mineralized and unmineralized fractures in the Yucca Formation.

Structures in Permian Host Rocks

At the Plata Verde mine, the main structural feature is a north-trending horst that formed during Basin and Range deformation (fig. 7). Elsewhere in the Van Horn Mountains, Twiss (1959a) observed east-northeast- and northeast-striking faults that terminate against, and therefore are presumably younger than, north- and north-northwest-striking Basin and Range faults. East-northeast and northeast faults occur in the southern part of the mine area (fig. 7), but their relative ages have not been established.

Chemical reduction of the Powwow red beds increases in intensity toward the south, where the faults are most closely spaced and where the Rim Rock Fault, a major Basin and Range fault with a throw of approximately 3,000 ft (910 m) (Twiss, 1959a) is closest to the Powwow beds (Price, 1982). Gray, reduced beds are more abundant in the south and red hematite-rich beds are more abundant in the north. Aqueous sulfide reductants may have been introduced into the Powwow sediments along the closely spaced faults. Geochemical zonation, which is thermodynamically predictable in its correlation with oxidation versus reduction, supports the hypothesis that reductants were epigenetically introduced along faults (fig. 7). Although the pods of secondary ore at the Plata Verde mine locally coincide with bedding, structural control involving east-northeast and northeast faults suggests that primary mineralization here may have been contemporaneous with that in Precambrian and Cretaceous rocks.

Changing Stress Orientations with Time and Probable Age of Vein Mineralization

Regional structural information indicates that stress orientations changed through time in the Trans-Pecos region (fig. 23). Maximum compression, σ_1, during Laramide deformation changed from northeast to east-northeast. Veins formed during early Laramide deformation would have been preferentially oriented northeast perpendicular to σ_3 and parallel to σ_1, whereas veins formed during late Laramide deformation would have been preferentially oriented east-northeast. Contemporaneous east-northeast and northeast veins would not have been likely, except perhaps during a period of Laramide deformation intermediate between the early and late stages. It is unlikely that significant vein mineralization, which requires flow of large volumes of water along fractures, could have occurred while Laramide compression was shortening the upper crustal rocks.

Stress orientations during middle Tertiary magmatism were essentially the same as those during late Laramide deformation, although compression had waned (Price and Henry, 1984). East-northeast- to east-striking dikes and igneous-related hydrothermal veins developed at this time (fig. 14), when σ_3 was north-northwest to north. The observation that east-northeast and northeast veins in red-bed sequences formed contemporaneously (σ_3 most likely being northwest rather than north-northwest or north) suggests that mineralization in red beds did not form during middle Tertiary magmatism.

Early Basin and Range extension would not have been conducive to ore deposition along east-northeast or northeast fractures. The direction favored for vein formation would have been north-northwest, perpendicular to σ_3 (fig. 23). The north-northwest-striking Rim Rock dikes formed at this time (fig.15).

Northwest-oriented extension during late Basin and Range deformation could have caused new fractures to develop (table 7) or could have caused preexisting east-northeast and northeast fractures to open. Silver-copper-lead veins in red-bed sequences of Trans-Pecos Texas most likely formed during this late Basin and Range extension. Major east-northeast fractures in the Hazel-Marvin-Judson-Mohawk area (fig. 2) may have originally formed during earlier deformation (perhaps as a sinistral set of en echelon tension fractures during late Laramide compression), but mineralization probably occurred during late Basin and Range extension. In the Indio Moun-

TABLE 7. Relative ages of fractures determined from fracture intersections in the Hazel Formation.

Number of intersections		Older trend ENE	NE	NW	WNW
	ENE		3*	7	2
Younger trend	**NE**	2*		8	11
	NW	4	0		4
	WNW	1	1	1	

*Fracture fillings at the Mohawk and Hazel mines indicate contemporaneous opening of the ENE and NE ore trends.

tains, east-northeast- and northeast-striking Laramide shear fractures were probably opened later during Basin and Range deformation.

By analogy with studies in New Mexico (Golombek, 1982), it is assumed that in Texas the late stage of Basin and Range extension began approximately 10 m.y.a. In Texas, silicic magmatism, with which numerous hydrothermal ore deposits are associated, had ended by 28 m.y.a., and volumetrically minor basaltic volcanism persisted to about 16 m.y.a. (McDowell, 1979). No younger igneous activity, with which less than 10-m.y.-old veins could be associated, is known.

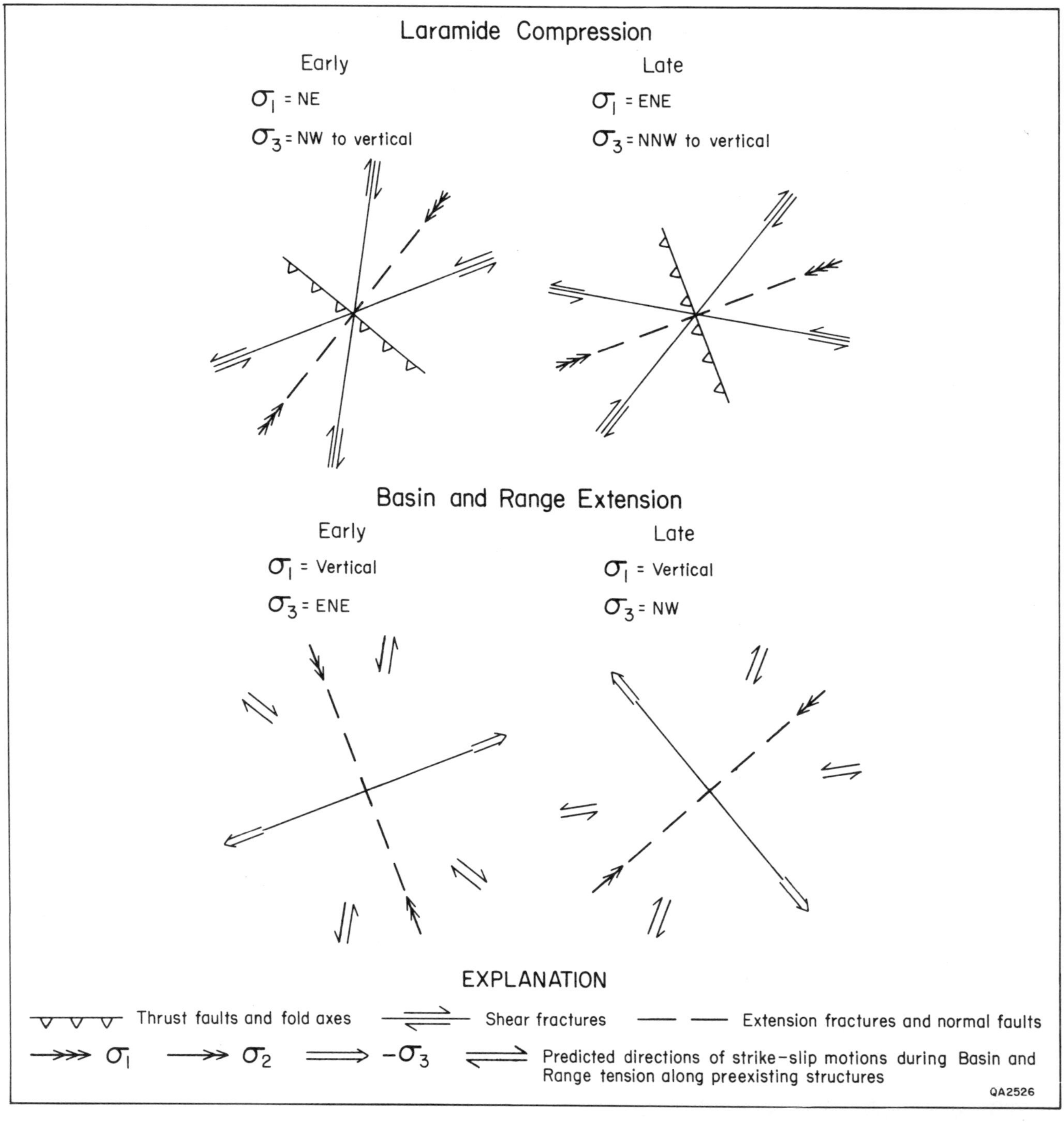

FIGURE 23. Proposed orientations of stress during Late Cretaceous through Recent(?) time, Trans-Pecos Texas. Stress orientations during middle Tertiary magmatism were probably similar to those during late Laramide compression.

Other Mineralization along Basin and Range Faults

Mineralization occurred along Basin and Range faults at several localities in the Trans-Pecos region. At the Mayfield manganese prospect (figs. 1 and 24), psilomelane (romanechite) and barite occur along the Mayfield Fault, an active Basin and Range fault (Twiss, 1959a; Price and others, 1983). The host rock, the Loma Plata Limestone, is silicified locally along the fault. Warren (1946) reported traces of zinc and lead at the prospect. Presumably aqueous barium, sulfide, and reduced manganese, moved upward along the fault. In the near-surface environment, the manganese and sulfur were oxidized and precipitated, perhaps because of dilution with shallow ground water.

King and Flawn (1953) noted a barite occurrence in the Carrizo Mountain Group along the Rim Rock Fault 2 mi (3 km) south of the Plata Verde mine. This fault is the western boundary fault of the Basin and Range horst that forms the Van Horn Mountains (fig. 1). Numerous other vein barite deposits in Trans-Pecos (Price and others, 1983b; Kyle, in preparation) may be associated with Basin and Range faults.

Beane (1974) and Putnam and others (1983) suggested that the galena-fluorite-barite deposits in the Hansonburg district, where mineralization occurs in Pennsylvanian limestone, and similar deposits elsewhere along the Rio Grande rift in New Mexico formed during Basin and Range deformation. Putnam and others (1983) indicated that (presumably pre-Basin and Range) northwest-striking, high-angle reverse faults served as conduits for the mineralization in the Hansonburg district. This orientation is most likely to have been extended during early Basin and Range deformation rather than during the later stage proposed for mineralization in red beds in Trans-Pecos Texas.

FIGURE 24. Open cut of manganese-barite vein along a Basin and Range normal fault at the Mayfield prospect. The host rock is the Lower Cretaceous Loma Plata Limestone.

MINERALOGY AND GEOCHEMISTRY OF ORES AND HOST ROCKS

Primary Ore Minerals

Primary sulfide minerals from veins in the Precambrian Hazel Formation indicate some geochemical variability between localities (table 8). The silver-copper ores at the Hazel mine (fig. 25) contain various silver and copper minerals, whereas the silver-copper-lead-zinc ores at the Pecos mine (fig. 26) contain dominantly copper-bearing sulfides plus galena and sphalerite. Because the high zinc content was a detriment to ore processing at the local smelter in

TABLE 8. Vein mineralogy, Precambrian Hazel Formation, Van Horn area.

	Vein mineralogy		Hazel mine	Marvin-Judson prospect	Mohawk mine	Eureka prospect	Pecos mine	Diablo prospect
Primary	chalcopyrite	$CuFeS_2$	x		x	x	x	
	tennantite-tetrahedrite	$Cu_3(As, Sb)S_{3+x}$	x				x	
	bornite	Cu_5FeS_4	x					
	galena	PbS					x	
	sphalerite	ZnS					x	
	pyrite	FeS_2	x	x	x		x	
	marcasite	FeS_2	x				x	
	acanthite (argentite)[1]	Ag_2S	x					
	barite	$BaSO_4$	x	x	x	x	x	
	calcite	$CaCO_3$	x	x	x	x	x	
Secondary	djurleite	$Cu_{1.93}S$	x					
	chalcocite[2]	Cu_2S	x					
	native silver[3]	Ag	x					
	covellite	CuS	x				x	
	malachite	$Cu_2CO_3(OH)_2$	x		x	x	x	x
	azurite	$Cu_3(CO_3)_2(OH)_2$					x	
	goethite	$FeOOH$	x	x	x	x	x	x

(Column heading over the locality columns: Locality)

[1]Argentite was reported by von Streeruwitz (1892b), who examined active mine workings.

[2]Chalcocite was identified petrographically by Flawn (1952). The chalcocite-appearing mineral found in near-surface mine workings during this study was identified by X-ray diffraction as djurleite. Some chalcocite could be primary.

[3]Native silver was reported by Flawn (1952) and von Streeruwitz (1892b). At the Hazel mine, Flawn also found an unidentified, strongly anisotropic brownish-gray mineral, which is paragenetically related to chalcocite, and an unidentified pale silver-white mineral.

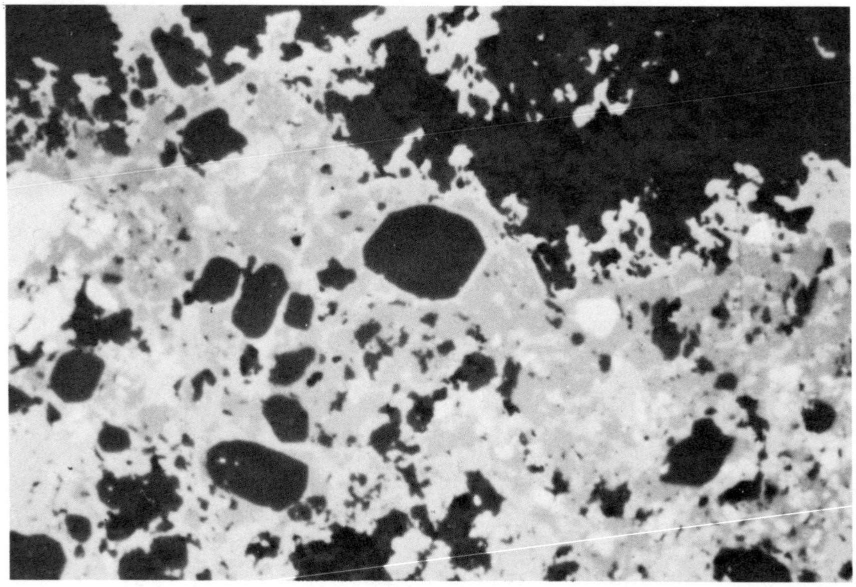

FIGURE 25. Photomicrograph of Hazel ore from glory hole, East shaft area. Long dimension of photograph is 1.1 mm, plane-polarized reflected light. Rounded sand and silt grains (dark gray) are surrounded by a mixture of barite (also dark gray), pyrite (white), djurleite (whitish gray), and bornite (gray).

El Paso, the Pecos vein was never extensively developed. Silver in the Pecos ore presumably is contained in the mineral tennantite.

Acanthite was identified by Price (1982) as a probable primary sulfide at the Plata Verde mine in the Permian Powwow red beds (table 9). Only oxide ores have been exploited at the mine. Exploration below the old workings, which reached a depth of 160 ft (49 m), would probably reveal primary sulfide mineralization.

Galena, chalcopyrite, and pyrite have been recognized as primary sulfides at deposits in the Cretaceous Yucca Formation in the Indio Mountains (table 10). Secondary zinc and arsenic minerals suggest the former presence of primary sphalerite and tennantite.

TABLE 9. Ore mineralogy, Powwow Member of the Permian Hueco Limestone, Plata Verde mine.

Mineral	Formula
bromargyrite	$Ag(Br,Cl)$
acanthite[1]	Ag_2S
malachite	$Cu_2CO_3(OH)_2$
azurite	$Cu_3(CO_3)_2(OH)_2$
chrysocolla	$CuSiO_3 \cdot nH_2O$
anglesite	$PbSO_4$
cerussite	$PbCO_3$
barite[2]	$BaSO_4$
limonite	FeOOH

[1]Acanthite is the only primary sulfide recognized at the Plata Verde mine. High copper, arsenic, lead, and zinc contents in the oxide ore suggest that primary sulfides included chalcopyrite, tennantite, galena, and sphalerite (Price, 1982).
[2]Calcite and manganese oxides occur in veinlets and along faults that are not obviously contemporaneous with ore deposition.

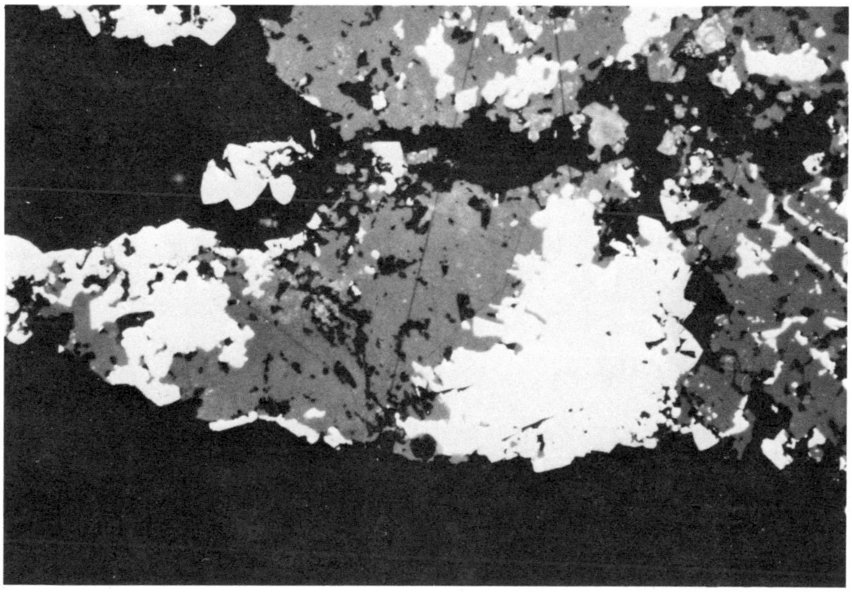

FIGURE 26. Photomicrograph of sample collected from the dump at the Pecos mine. Long dimension of photograph is 0.56 mm, plane-polarized reflected light. Barite (black) is interfingered with sulfides: sphalerite (gray) and galena, chalcopyrite, and tennantite (all white).

TABLE 10. Mineralogy, hosts, and controls on silver-copper-lead deposits, Cretaceous Yucca Formation, Indio Mountains.

	Vein mineralogy	Locality	Road prospect	Finlay prospect	Galena occurrence	Rossman prospect	Southwestern workings	North pit	Black Diamond mine	South pit	Carpenter prospect	Purple Sage mine
Primary	galena	PbS	x	x	x		x	x	x			x
	chalcopyrite	$CuFeS_2$				x						
	pyrite	FeS_2	x	x		x						
	barite	$BaSO_4$	x	x	x	x	x	x				
	calcite	$CaCO_3$	x	x		x	x	x	x	x	x	x
Secondary	djurleite	$Cu_{1.93}S$								x		
	low chalcocite	Cu_2S				x						
	covellite	CuS				x						x
	malachite	$Cu_2CO_3(OH)_2$	x	x		x	x	x	x	x	x	x
	azurite	$Cu_3(CO_3)_2(OH)_2$	x									
	chrysocolla	$CuSiO_3 \cdot nH_2O$	x			x	x		x		x	x
	brochantite	$Cu_4SO_4(OH)_6$								x		
	anglesite	$PbSO_4$	x	x			x					
	cerussite	$PbCO_3$										x
	wulfenite	$PbMoO_4$										x
	mimetite	$Pb_5(AsO_4)_3Cl$										x
	pyromorphite	$Pb_5(PO_4)_3Cl$							x			
	vanadinite	$Pb_5(VO_4)_3Cl$										x
	mottramite	$Pb(Cu,Zn)VO_4OH$										x
	plumbojarosite	$PbFe_6(SO_4)_4(OH)_{12}$		x								
	beudantite-group	$Pb_{1-x}(Fe,Al)_{3-y}(SO_4,AsO_4,SiO_4)_2(OH)_6$		x								
	hemimorphite	$Zn_4Si_2O_7(OH)_2 \cdot H_2O$										x
	goethite	FeOOH	x	x		x	x	x	x	x	x	x
	hematite	Fe_2O_3				x	x		x			
	manganese oxide	MnO_2	x			x			x			x

Hosts and controls on mineralization	Road prospect	Finlay prospect	Galena occurrence	Rossman prospect	Southwestern workings	North pit	Black Diamond mine	South pit	Carpenter prospect	Purple Sage mine
upper Yucca sandstone member (Y4)	x	x	x							
middle Yucca sandstone-conglomerate member (Y3)					x	x	x			x
middle Yucca conglomerate-sandstone member (Y2)				x				x	x	
lower Yucca conglomerate member (Y1)				x						
veins along high-angle faults and fractures	x	x	x	x	x	x	x	x		
disseminated mineralization following bedding				x					x	x

Secondary Minerals

Numerous secondary minerals have been recognized at different localities in red beds (tables 8, 9, and 10). Particularly noteworthy are bromargyrite, the main silver mineral at the Plata Verde mine (Price, 1982), and several unusual oxide minerals in the Indio Mountains (table 10 and fig. 27). An arsenic-rich sample of outcropping vein from the Finlay prospect proved to be a mixture of barite, quartz, and a sulfur-rich beudantite-group mineral (tables 11 and 12). The beudantite-group mineral occurs as light-orange massive aggregates of grains and as hexagonal platelets generally less than 0.015 mm wide and 0.003 mm thick. This previously undescribed mineral, which is in the jarosite group as defined by Botinelly (1976), is arsenic-poor and sulfur-rich relative to beudantite, which has the ideal formula $PbFe_3SO_4AsO_4(OH)_6$.

FIGURE 27. Photomicrograph of clusters of radiating hemimorphite blades, Purple Sage mine, Indio Mountains. Magnification is 15X.

TABLE 11. X-ray powder diffraction data on a sulfur-rich beudantite-group mineral, Finlay prospect, Indio Mountains.

Relative intensity	Indices hkl[1]	Measured d-spacing[2], Å	Calculated d-spacing[3], Å
6	101	5.94	5.93
55	003	5.64	5.65
5	110	3.65	3.66
15	104	3.50	3.52
14	113	3.07	3.07
3	202	2.96	2.97
100+	006	2.83	2.83
4	024	2.54	2.54
24	107	2.27	2.26
3	116	2.24	2.24
3	033	1.977	1.977

[1]Using beudantite structure ($R\bar{3}m$) for indexing (Walenta, 1966)
[2]Cu-K_{α} radiation; quartz as internal standard
[3]Using a = 7.31 Å, c = 16.95 Å

TABLE 12. Electron-microprobe analysis of a sulfur-rich beudantite-group mineral, Finlay prospect, Indio Mountains.

Element	Weight percent ± 1σ (4 analyses)
Pb	28.87 ± 0.38
Fe	22.22 ± 0.41
Al	0.39 ± 0.11
As	5.61 ± 0.83
S	6.66 ± 0.40
Si	0.38 ± 0.13
structural O[1]	32.09 ± 0.33
structural H[1]	0.87 ± 0.01
Total[2]	97.09 ± 0.74

[1]Calculated on the basis of a structural formula assuming (a) limited substitution of Si for charge balance and (b) S + Al + Si =2:
$(Pb_{.97}\square_{.03})(Fe_{.93}Al_{.03}\square_{.04})_3[(S_{.73}As_{.26}Si_{.01})O_4]_2(OH)_6$, where □ is a vacancy. Excess Si is contamination by quartz.
[2]Energy-dispersive spectra and trace-element analysis of bulk sample 82-37 containing beudantite, quartz, and barite (see table A5 in the appendix) indicate no other elements that are in concentrations high enough to affect the beudantite formula given above.

Gangue Minerals

Barite and calcite are the dominant gangue minerals in veins in Precambrian and Cretaceous rocks. Calcite and barite also occur along faults at the Plata Verde mine, but their relation to the secondary ores is unclear. Textural features suggest that barite and calcite were deposited in veins more or less contemporaneously with primary sulfides. Samples from the dump at the Pecos mine display alternating bands of barite and primary sulfides and local complex interfingering of barite and sulfides (fig. 26). At the Hazel mine, Flawn (1952) noted breccias containing sandstone fragments having successive coatings of sulfides, barite, and more sulfides.

Unlike igneous-related hydrothermal veins in Trans-Pecos Texas, quartz is not a major gangue mineral. Flawn (1952) indicated that quartz in the Hazel ores is dominantly detrital. Other detrital or diagenetic components in the sandstones include (a) *in the Precambrian Hazel Formation,* microcline, orthoclase, plagioclase, muscovite or illite, chlorite, biotite, tourmaline, green amphibole, zircon, montmorillonite, sedimentary rock fragments (chert and dolomite), and metamorphic rock fragments (mica schist); (b) *derived from the Carrizo Mountain Group in the Permian Powwow red beds,* microcline, muscovite, plagioclase, chlorite, biotite, and tourmaline; and (c) *in the Cretaceous Yucca Formation,* sedimentary rock fragments (mostly chert, shale, and limestone), muscovite, tourmaline, zircon, biotite, and trace amounts of alkali feldspar. The Yucca sediments had sources different from those of the Permian and Precambrian red beds. Some of the detrital quartz in the Yucca Formation, however, probably had an igneous-hydrothermal origin. Some detrital quartz grains contain halite-bearing, three-phase fluid inclusions, which are typical of high-temperature igneous-hydrothermal sources. Calcite and hematite locally cement sandstones in all three formations, and dolomite is a cement in the Hazel Formation.

Quartz, a locally minor gangue mineral in veins in Precambrian and Cretaceous rocks, is probably remobilized detrital silica that reprecipitated during vein formation. At the Hazel mine, quartz occurs locally with barite in the matrix between fragments of brecciated sandstone. At the Finlay prospect in the Indio Mountains, detrital quartz grains are overgrown with quartz that protrudes into a galena-barite vein (fig. 28).

Fluorite, a mineral commonly found in igneous-hydrothermal veins in Trans-Pecos Texas (Price and others, 1983b), has not been identified in any of the veins in red-bed sequences. The low quartz contents and common lack of fluorite in these veins are further evidence that the silver-copper-lead mineralization in red beds is unrelated to middle Tertiary igneous activity.

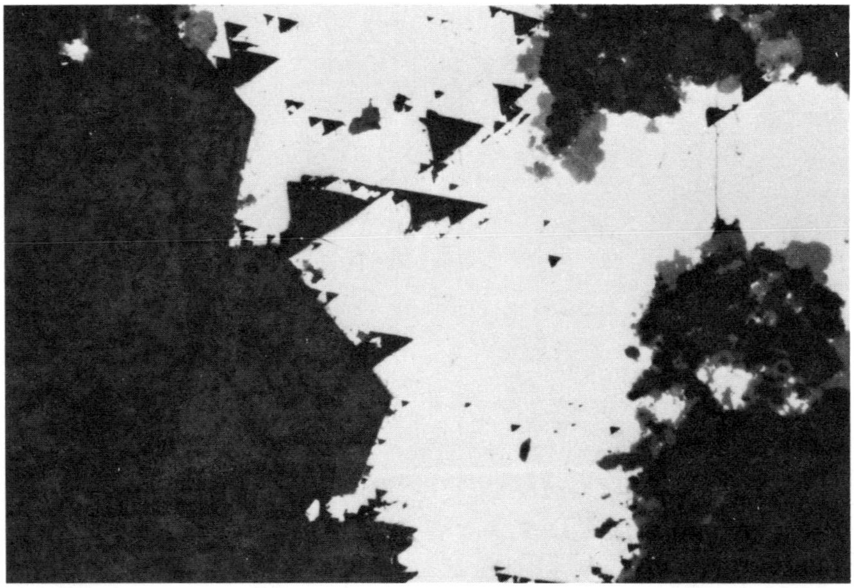

FIGURE 28. Photomicrograph of quartz overgrowths, Finlay prospect, Indio Mountains. Long dimension of photograph is 1.1 mm, plane-polarized reflected light. Along the edge of a galena (white) - barite (light and dark gray) vein, detrital quartz grains are overgrown by quartz having hexagonal crystal outlines. Triangular black areas are polishing pits in galena.

Wall-Rock Alteration

Wall-rock alteration, typically intense near igneous-hydrothermal veins, is minimal near the silver-copper-lead veins cutting red beds. Much of the apparent alteration is probably due to secondary or supergene processes. Weathering of sulfides in the veins produced enough acid to bleach the red wall rocks in certain areas. Local orange discoloration is the result of limonitic staining of the wall rocks. Flawn (1952) recognized that wall rocks at the Hazel mine in places contain veinlets of calcite and barite and that some primary alteration and discoloration of the red beds did occur. The characteristically red sandstone of the Hazel Formation is locally gray (hematite removed) and is impregnated with sulfides. Similar alteration at the Pecos mine involves pyrite-marcasite veinlets in gray sandstone (fig. 29a).

Detrital minerals that would be expected to be unstable with respect to the acidic ore-forming fluids, such as plagioclase, microcline, muscovite, and biotite, are apparently unaffected by the primary alteration near the veins. At the Pecos mine, for example, unaltered microcline occurs in contact with vein-related marcasite and pyrite in the gray, weakly altered sandstone (fig. 29b). Geochemical data (tables B1 through B8 in the appendix) indicate decreased sodium content of limonitic wall rock next to veins (table B2) relative to unaltered country rock (table B3), a change that may be due to acid leaching associated with sulfide weathering. Some plagioclase destruction may have accompanied primary alteration; a similar drop in sodium content was detected in one of the two Pecos mine samples analyzed (table B2, sample 82-31).

Price (1982) noted a lack of hydrothermal alteration at the Plata Verde mine, where detrital microcline, plagioclase, and muscovite appear to have been unaffected by the ore-forming fluids. Variation in the color of the Powwow sediments may be considered a form of alteration resulting

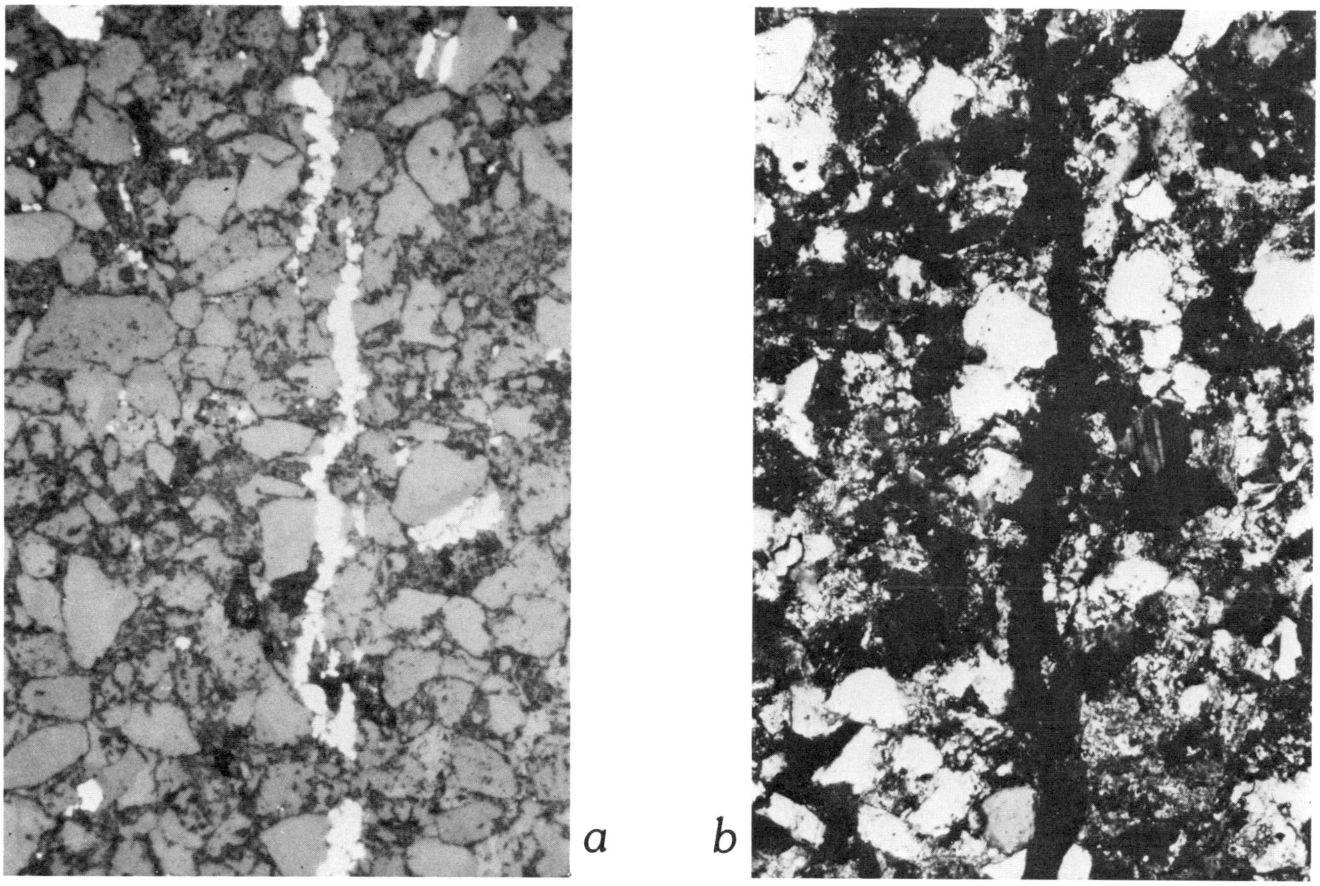

FIGURE 29. Photomicrographs of gray, weakly altered sandstone, Pecos mine. Long dimension of photographs is 1.1 mm. (a) Plane-polarized reflected light: pyrite-marcasite veinlet and impregnations in sandstone; (b) same view in transmitted light with crossed polarizers. Note twinned detrital microcline grain next to pyrite-marcasite mass near center of photograph and veinlet trending from top to bottom.

from syngenetic combined with epigenetic reduction. As discussed previously (on page 29), the epigenetic reduction is probably related to closely spaced faults in the southern part of the mine area (fig. 7). Ore occurs in reduced, gray rocks and in formerly reduced, but now limonitic orange rocks, rather than in oxidized red beds themselves.

In the Indio Mountains, clay minerals occur locally along the faults that were extended and filled with vein material. At the Road prospect (fig. 8), montmorillonite was identified by X-ray diffraction. At the Southwestern workings, which may be along the same fault as the Road prospect, sericite (muscovite) and a kaolinite-group mineral were identified. The clay-mineral zones are not extensive at any of the localities; the fault zones in which the clays occur are generally less than 6 ft (2 m) wide, and many of the veins have no clay zones at all. It is not clear whether the clay minerals formed at the same time as the silver-copper-lead-barite mineralization, much earlier during Laramide shearing, or much later during weathering of the sulfides.

Mineralizing solutions apparently penetrated the permeable sedimentary beds away from some of the veins. Anomalous concentrations of silver are present several hundred feet from likely major fracture conduits for mineralizing fluids at the Plata Verde mine (Price, 1982). Geochemical data (tables B2 and B3 in the appendix) reveal that primary dispersion halos generally extend no more than 10 ft (3 m) beyond the veins in the Hazel Formation. Silver tends to be more widely dispersed than copper, lead, zinc, or arsenic (table B2). At several localities, red sandstones and siltstones collected from within 10 ft (3 m) of the veins exhibit no apparent mineralogical or geochemical effects from the nearby veins (table B3).

In the Indio Mountains, disseminated ore minerals locally occur far from obvious faults or veins (table 10). At the northernmost Rossman prospect pit (fig. 8), for example, chalcocite, barite, and minor amounts of chalcopyrite occur as cement and possibly as replacements of limestone rock fragments in Yucca sandstone (fig. 30). As far as 50 ft (15 m) from the Finlay vein, presumably primary barite occurs as pore fillings in sandstone. Isolated specks of limonite in the sandstone at this locality suggest the former presence of pyrite. Anomalously large amounts of lead and arsenic were detected 95 ft (29 m) from this vein (table B6). In general, however, primary geochemical halos are much more restricted to the immediate vicinity of the veins.

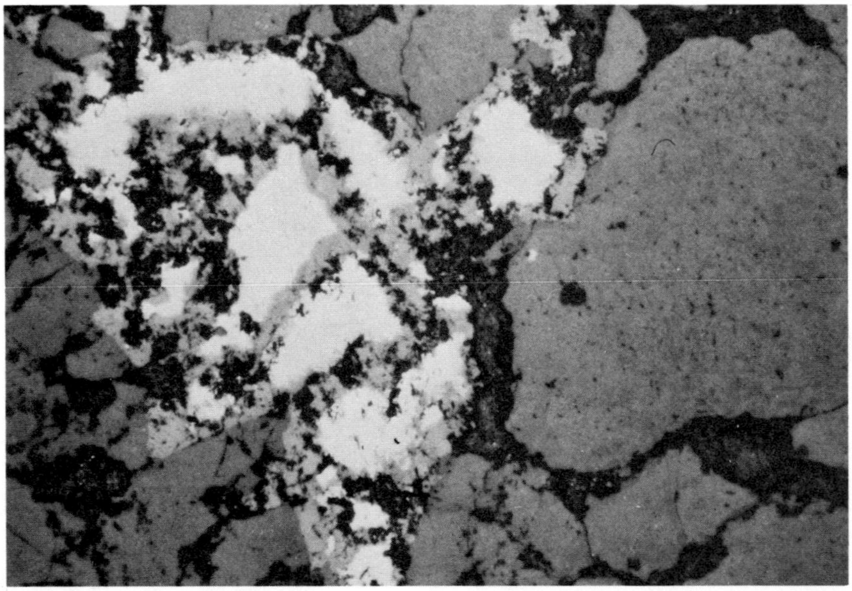

FIGURE 30. Photomicrograph of disseminated copper mineralization at the Rossman prospect, Indio Mountains. Long dimension of photograph is 0.56 mm, plane-polarized reflected light. Chalcocite (white), barite (dark gray), and secondary malachite (light gray) cement quartz (medium gray) sand grains. The large mass of chalcocite may be the replacement of a dissolved limestone fragment. Minor amounts of chalcopyrite are surrounded by chalcocite in this rock. Chalcocite may be supergene in origin, but chalcopyrite is probably primary.

Geochemical Character of the Ores

Ores from the three host formations are geochemically similar, as would be expected if they share a common origin. Geochemical data on samples in Precambrian and Cretaceous rocks are listed in this report as tables B1 through B8 in the appendix. Data on Permian rocks were tabulated by Price (1982), who also outlined the analytical procedures used: inductively coupled plasma-atomic emission spectroscopy for all elements except silver (flame atomic absorption) and gold (graphite-furnace atomic absorption). Geochemical data on ore samples were then compared with data on unmineralized sandstones from the same formations (figs. 31 through 33).

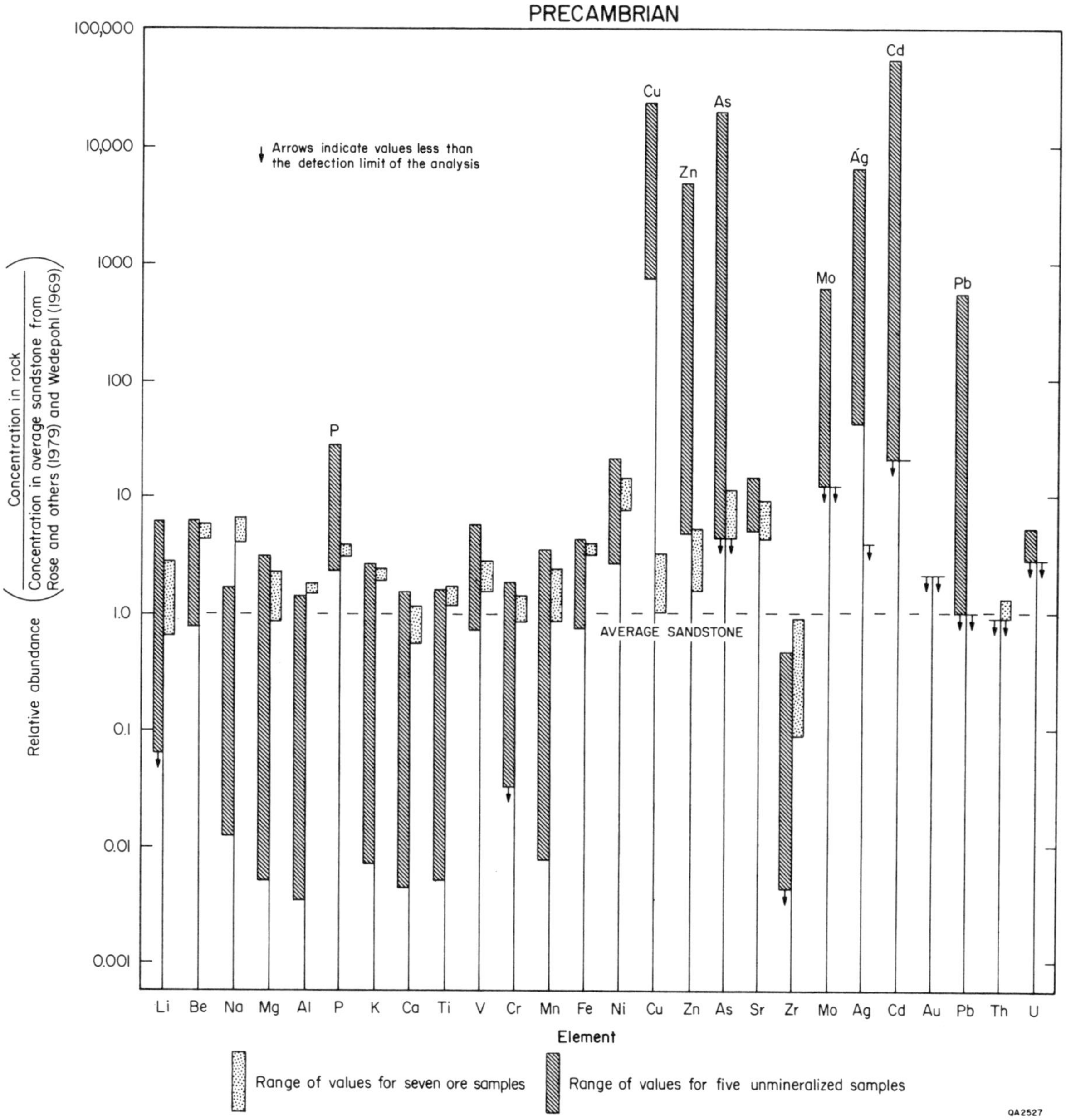

FIGURE 31. Geochemical characteristics of ores and unmineralized rocks in the Precambrian Hazel Formation.

Each of the three formations shows significant enrichments in copper, silver, lead, arsenic, zinc, cadmium, and molybdenum. Barium and sulfur, which were not analyzed, probably contain strong enrichments, as evidenced by the common occurrence of barite as a gangue mineral. Phosphorus, vanadium, and uranium are locally enriched. Apparent depletions shown in figure 31, which incorporates the high-grade samples listed in table B1, indicate that the elements lithium, beryllium, sodium, magnesium, aluminum, potassium, titanium, chromium, manganese, and zirconium were probably not deposited by the ore-forming fluids. Calcium is locally depleted and enriched, occurring as minor amounts of calcite gangue in veins.

Comparison with Igneous-Hydrothermal Ores

The lack of gold enrichment in these ores is noteworthy. Silver-to-gold ratios of the ores are high (table 1) relative to typical igneous-hydrothermal ores. Price (1982) argued that the high silver-to-gold ratio of the Plata Verde ore (87,000 to 1) was consistent with relatively low-temperature, ore-forming fluids unassociated with igneous activity. Although traces of gold were reported by von Streeruwitz (1892b) at the Hazel mine, no gold is known to have been produced from any of the mines in the Hazel Formation. Geochemical analyses of high-grade silver samples indicate silver-to-gold ratios greater than 160,000 to 1. Gold has not been detected in any of the

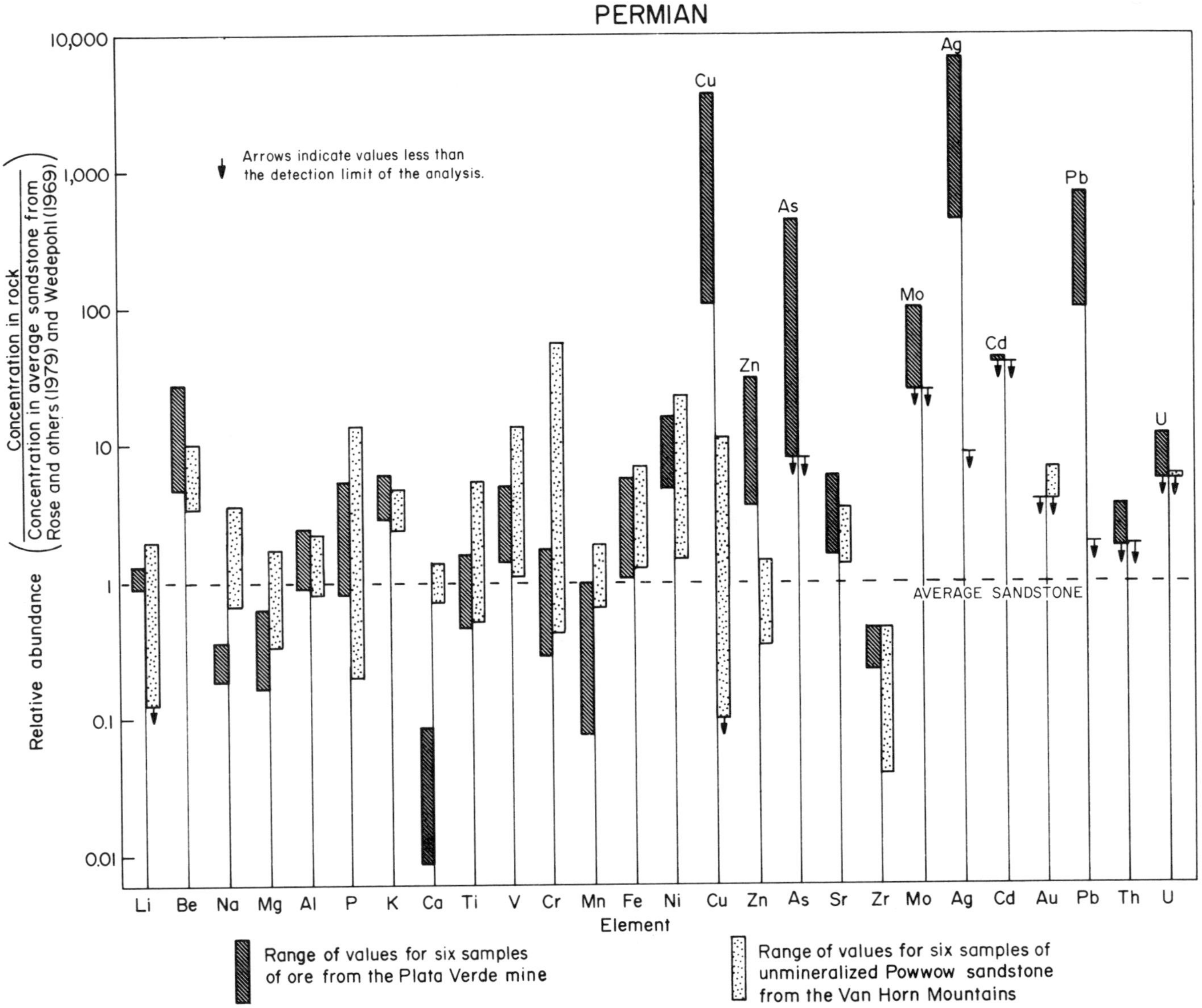

FIGURE 32. Geochemical characteristics of Plata Verde ores and unmineralized rocks in the Powwow Member of the Permian Hueco Limestone (from Price, 1982).

samples from any of the formations yet analyzed (limit of detection is 0.01 parts per million by weight).

Igneous-hydrothermal ores in the Trans-Pecos region, where igneous rocks are alkalic and alkali-calcic, are typically high in silver-to-gold ratios relative to ores associated with calc-alkaline igneous rocks (table 13). Epithermal gold-silver ores associated with calc-alkaline rocks have silver-to-gold ratios in the range of 0.1 to 100 (Berger and Eimon, 1983). The Trans-Pecos igneous-hydrothermal ores are characteristically low in gold, but their silver-to-gold ratios are nevertheless generally an order of magnitude lower than the extremely gold-poor silver-copper-lead deposits in red beds. Because gold is less mobile than silver, copper, and lead at low temperatures, it is likely that the ores in red beds formed at lower temperatures than did the igneous-hydrothermal veins.

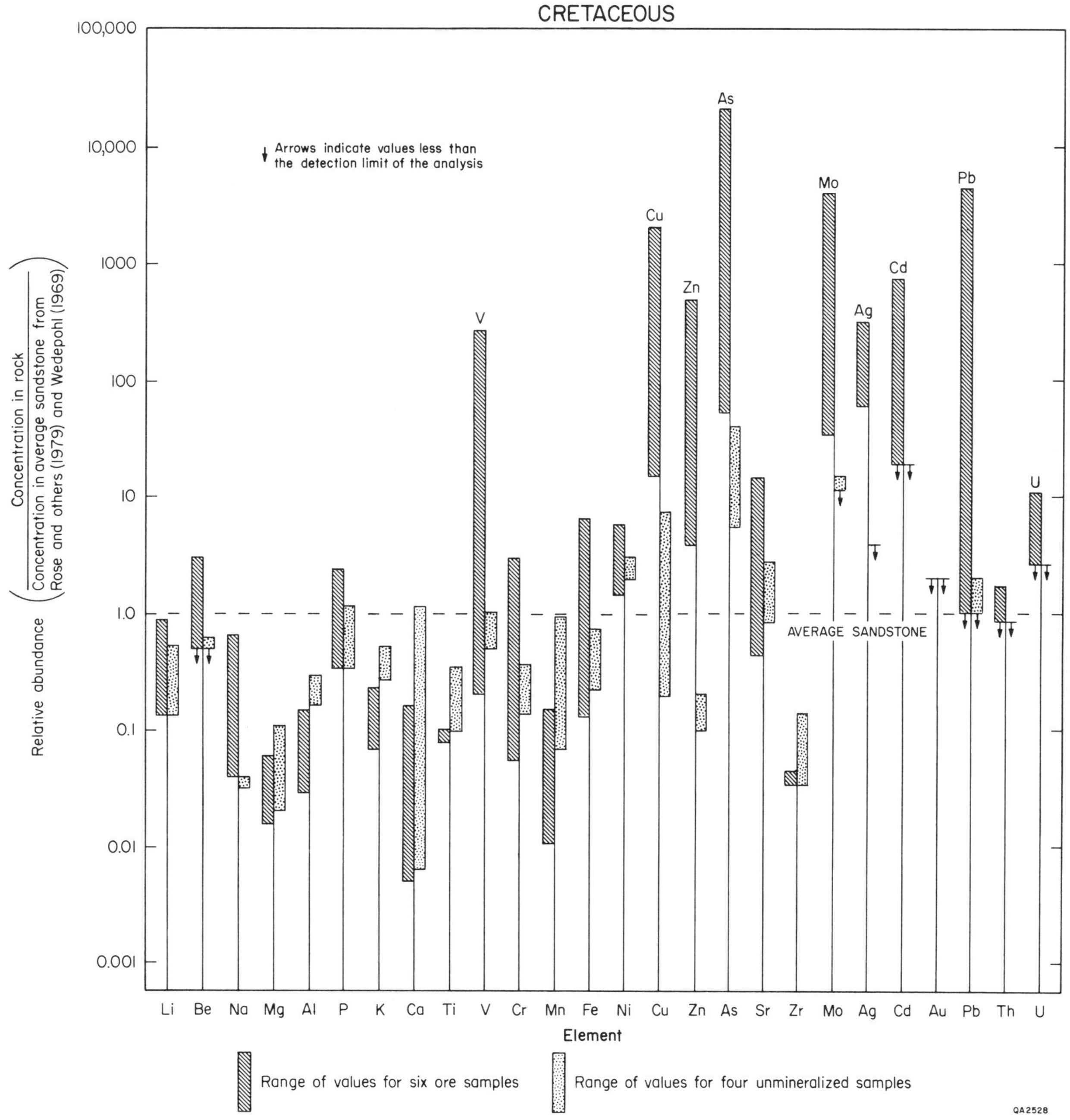

FIGURE 33. Geochemical characteristics of mineralized and unmineralized rocks in the Cretaceous Yucca Formation.

TABLE 13. Silver grades and silver-to-gold ratios in middle Tertiary, igneous-hydrothermal ores in the Trans-Pecos region.

Locality	Sample	Silver grade (oz/ton)	Ag/Au (by weight)	Reference
Chinati Mountains area, Presidio County				
Silver-lead mantos, Shafter District				
Presidio mine	~1,000,000 tons	13	2,300	Ross (1943)
Perry mine	<100 tons	6	120	McMillan (1949)
Sullivan mine	>4,000 tons	5	65	Ross (1943)
Epithermal lead-zinc-silver veins				
San Antonio mine	1 ore sample	0.5	>1,800	this report
Burney mine	2 ore samples	1.0; 2.7	>3,500; >9,200	this report
Wood prospect	10.8-ft core	20.1	690	J. V. Bikun and F. D. Busche (personal communication, 1984)
Sierra Rica caldera complex, Chihuahua, Mexico				
Lead-zinc-silver skarn, San Carlos mine	~1,600,000 tons	3	390	Immitt (1981)
Quitman Mountains area, Hudspeth County				
Epithermal lead-zinc-silver veins				
Bonanza mine	1 ore sample	0.2	315	this report
Tarantula Hills area	~39 tons	3.5	12	Henderson and Mote (1945)
Lead-zinc-silver-molybdenum skarn				
Zimpleman Pass area	1 ore sample	4.7	11,000	this report
Eagle Mountains, Hudspeth County				
Epithermal lead-zinc-silver veins				
Black Hill mine	6 ore samples	7.8 to 47 (avg. 17)	54 to 3,100 (avg. 250)	this report
Altuda area, Brewster County				
Lead-silver breccia pipe				
Bird mine	3 ore samples	1.3 to 7.8	5,300 to >9,600	this report

Comparison with Other Ores in Red Beds

Price (1982) compared the geochemical characteristics of the Plata Verde ores with those of strata-bound ores in red beds: copper shales, cupriferous sandstones, and uraniferous sandstones. Many geochemical similarities exist, such as low gold contents and local enrichments in elements found in the Trans-Pecos ores (figs. 31 through 33). In contrast, cupriferous sandstones typically have higher copper-to-silver ratios (generally greater than 40 to 1 [Price, 1982]) than do the Trans-Pecos ores (table 1). Structural control of the Trans-Pecos ores clearly points to an origin different from that of red-bed copper ores. Fluid-inclusion studies described in the following section indicate temperatures of vein formation higher than those usually attributed to syngenetic-diagenetic strata-bound copper deposits.

Fluid-Inclusion Studies

Samples

Fluid inclusions in vein minerals from the Pecos mine (Precambrian Hazel Formation) and the Finlay prospect (Cretaceous Yucca Formation) reveal aspects of the physical and chemical conditions of ore formation. Two-phase aqueous inclusions are common in barite, calcite, and sphalerite from the Pecos mine (fig. 34). All three minerals probably precipitated from the same ore-forming fluid because (1) the estimated ratios of bubble volume to inclusion volume are similar in all three minerals and (2) the banded ore samples, in which barite-calcite-sulfide bands alternate with sulfide bands, suggest that all three minerals formed more or less contemporaneously. Measurements of heating and freezing temperatures of large (2 to 20 μm) inclusions in barite were made using the Fluid, Inc., apparatus owned by the Department of Geological Sciences at The University of Texas at Austin, but inclusions in calcite and sphalerite were too small (<1 μm) for microthermometry. Measurements were made on inclusions in calcite and barite from the Finlay prospect. Similar-appearing inclusions in quartz were too small for measurements of heating and freezing.

Data from Heating

The homogenization temperature of a fluid inclusion is a function, among other things, of the temperature at which the crystal trapped the inclusion (Roedder, 1979). The ore-forming fluid

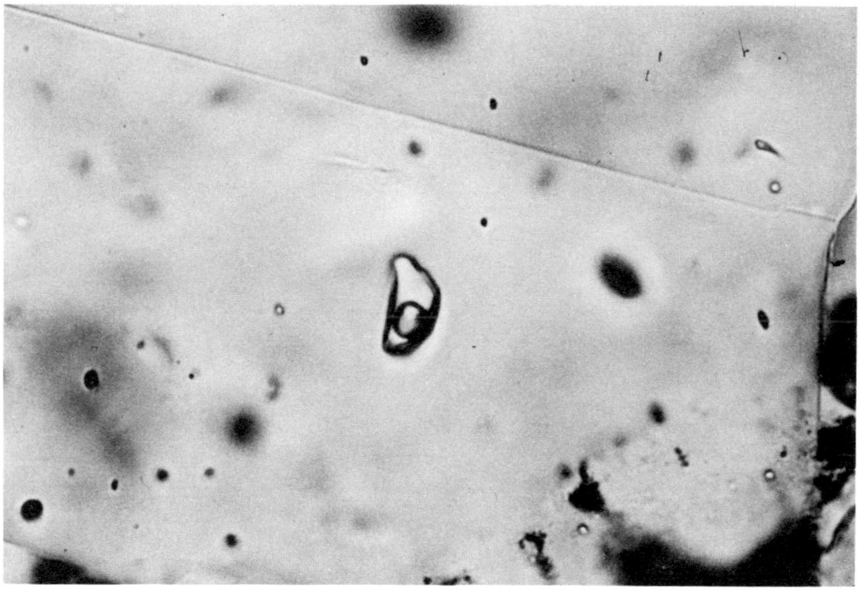

FIGURE 34. Two-phase (liquid-vapor) fluid inclusion in barite, Pecos mine. Note the cleavage trace in this transmitted-light photograph. Vapor bubble is 0.004 mm in diameter. Inclusions having similar vapor-to-liquid ratios occur in sphalerite and calcite within the same sample. Upon heating, vapor bubbles in inclusions disappear or homogenize into the liquid phase generally between 140°C and 170°C. Freezing temperatures of primary inclusions suggest that the ore-forming fluids contained from 9 to 19 weight percent equivalent NaCl.

was presumably a one-phase fluid, which, after being trapped in the crystal and upon cooling and lowering of pressure, exsolved a vapor bubble. Experimental heating reverses the natural cooling process and, ideally, homogenization occurs at a temperature somewhat lower than the temperature of ore formation. If a significant thickness of overburden capped the vein at the time of inclusion entrapment, a pressure correction is required to adjust the homogenization temperature to the higher formation temperature.

Both primary and secondary fluid inclusions were examined (fig. 35). Inclusions that were isolated and not obviously aligned along fractures were judged to be primary or to contain a sample of the ore-forming fluid. Secondary inclusions that were aligned along fractures probably represent younger, retrograde fluids and usually yielded lower homogenization temperatures than did primary inclusions.

A general assumption in interpreting fluid-inclusion data is that the volume of the inclusion remains constant despite changes in temperature. Because freezing of liquid-rich inclusions can expand the walls of the inclusions the heating experiments were always run before the freezing experiments. The assumption of volume constancy is generally valid for minerals without good cleavage, such as quartz, but may be invalid for minerals that cleave or deform easily, such as

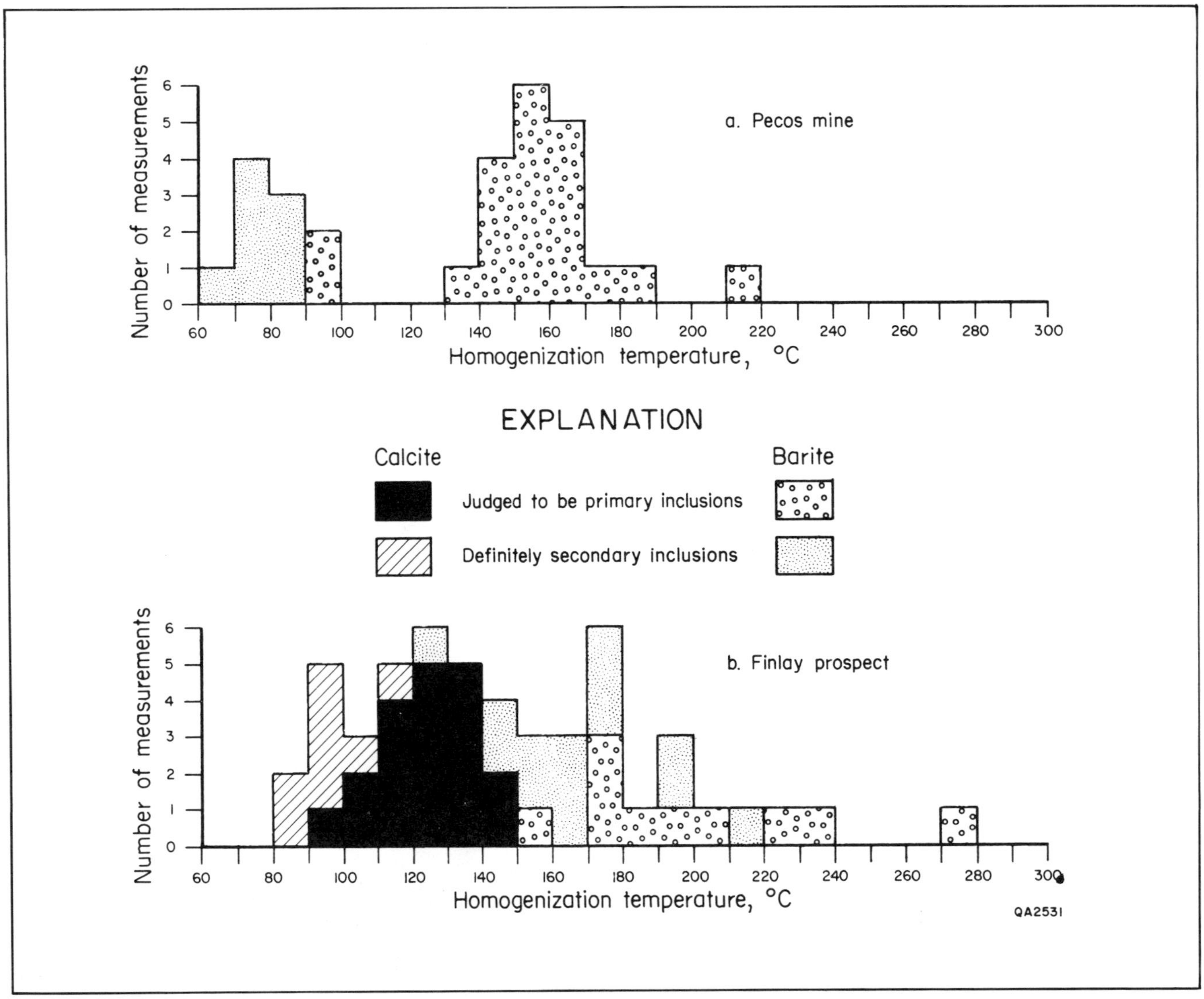

FIGURE 35. Homogenization temperatures of fluid inclusions from (a) the Pecos mine in the Precambrian Hazel Formation and (b) the Finlay prospect in the Cretaceous Yucca Formation. Ores probably formed in the temperature range of 120°C to 170°C.

barite. Unless heated well above their homogenization temperatures, inclusions in calcite generally maintain their volume (M. R. Ulrich, personal communication, 1983). Heating barite can raise the pressure inside the inclusion high enough to expand the volume of the inclusion, thus raising the homogenization temperature and yielding an erroneously high formation temperature (Ulrich and Bodnar, in press).

Some inclusions in barite expanded during the heating experiments and in some, visible cleavage traces opened slightly. In other fluid inclusions, as homogenization was approached with continued heating, the gas bubble decreased in size, as expected, then increased slightly (when the inclusion expanded), then decreased again before final homogenization into the liquid phase. Inclusions in barite therefore yield potentially high homogenization temperatures.

Homogenization temperatures of primary fluid inclusions in calcite from the Finlay prospect (fig. 35) suggest ore formation at temperatures between 120°C and 150°C. A pressure correction of much more than 9°C is unlikely to be necessary. This value is the correction corresponding to a 3,200-ft (970-m) column of water having a density of 1.1 g cm^{-3}. Data from Lemmlein and Klevtsov (1961) indicate that for water containing 15 weight percent NaCl (see page 46) and homogenization in this range, the change in temperature beyond homogenization with respect to a change in pressure at constant volume is approximately 0.084°C bar^{-1}. Stratigraphic reconstruction is useless in determining thickness of overburden at the Finlay prospect because Laramide thrusting and folding disrupted the layering. At the time the Eagle Mountains caldera erupted, 37 m.y.a. (Henry and Price, 1984), Cretaceous rocks formed the caldera walls. The elevation difference between the highest point in the caldera and the Finlay prospect is 3,200 ft (970 m). Whether this much rock covered the prospect during mineralization is uncertain.

Homogenization temperatures of inclusions in barite from the Finlay prospect vary considerably and are generally higher than those in calcite (fig. 35b). This scatter in values and shift to higher temperatures probably resulted from expansion of the inclusions during heating experiments. Inclusions that expanded greatly yielded the highest homogenization temperatures. Assuming that the barite and calcite originally precipitated at the same temperature, the average shift caused by inclusion expansion is 70°C to 80°C.

Inclusions in barite from the Pecos mine have more uniform homogenization temperatures than do those from the Finlay prospect (fig. 35). It is possible that the Pecos mine inclusions expanded less severely upon heating than did the Finlay prospect inclusions, perhaps because Basin and Range faulting was more intense in the Indio Mountains than near the Pecos mine. The 140°C to 170°C peak in homogenization temperatures may correspond to the actual temperature of ore formation at the Pecos mine. Alternatively, if expansion did significantly raise the homogenization temperatures, 170°C can be considered a probable maximum temperature of ore deposition.

Stratigraphic reconstruction at the Pecos mine indicates that a pressure correction to adjust the homogenization temperature to the formation temperature is minor. A likely minimum thickness of overburden is 1,500 ft (450 m), the difference in elevation between the Pecos mine and the top of the nearby Permian limestone. No one knows exactly how much of the Cretaceous sedimentary and Tertiary volcanic sections capped the Permian limestone here at the time of ore deposition: King (1965) estimated Cretaceous rocks to be approximately 2,500 ft (770 m) thick in the Sierra Diablo region, and Twiss (1959a) indicated a thickness of 3,600 ft (1,090 m) in the Van Horn Mountains. It is unlikely that more than 400 ft (120 m) of Tertiary tuffaceous sediment and volcanic rock covered the southern Sierra Diablo. The hydrostatic pressure of a column of hydrothermal fluid filling a fracture through the maximum thickness of overburden, 5,500 ft (1,660 m), would require a correction of only 15°C to the homogenization temperature.

The temperature range of ore deposition suggested by the fluid inclusion studies, 120°C to 170°C, is low relative to typical igneous-hydrothermal veins that contain copper, lead, and silver. Similar temperatures were, however, determined in fluids that formed lead-fluorite-barite deposits in the Rio Grande rift region of New Mexico (Beane, 1974; Putnam and others, 1983). Deposits in New Mexico formed during rifting, and the heat source for the fluids was either shallow igneous intrusions or the high heat flow associated with the Rio Grande rift, a part of the Basin and Range province (Beane, 1974; Putnam and others, 1983). The presence of numerous hot springs along the Rio Grande in Chihuahua and Texas, where no igneous activity younger than 16 m.y.a. is known (McDowell, 1979), indicate that even today hot waters flow upward along Basin and Range faults

(Henry, 1979); water temperatures as high as 90°C at the surface were reported by Henry (1979).

Data from Freezing

The freezing point of water is lowered by dissolved components. The dominant dissolved component is sodium chloride (NaCl), and freezing-point depressions measured by microthermometry are generally expressed in terms of salinities or as weight percents of equivalent amounts of dissolved NaCl (fig. 36). Primary fluid inclusions from the Pecos mine indicate that ore-forming fluids were saline (averaging 14 weight percent equivalent NaCl) but variable (9 to 19 percent). The high salinities suggest that the ore-forming fluids originated at depth and moved upward along the veins. Downward-migrating meteoric waters are thought to have been less saline. Putnam and others (1983) reported comparable salinities in lead-fluorite-barite deposits along the Rio Grande rift.

The variable salinities suggest that the chemical composition of the fluids changed drastically during ore formation, perhaps because rising saline fluids were diluted by shallow ground water. Although boiling as a result of drop in pressure upon ascent of the hydrothermal fluids also could have produced the observed differences in salinity, independent evidence of boiling, such as variable vapor-to-bubble ratios, is lacking. Solubility of the major metals, copper, silver, lead, and zinc, is enhanced by high salinity. Mineral precipitation was probably caused by decomplexing, other chemical changes, or temperature drop resulting from mixing the two fluids. Younger fluids trapped in secondary inclusions have lower salinities (fig. 36) and lower homogenization temperatures (fig. 35) and are probably samples of diluted, retrograde ore-forming fluids or relatively fresh ground water.

Comparison with Other Areas of Known Hydrothermal Activity Associated with Basin and Range Faulting

The manganese oxide (romanechite)-barite-quartz mineralization at the Mayfield prospect (Price and others, 1983b) follows the Mayfield Fault, a Basin and Range normal fault that forms the eastern boundary of the Van Horn Mountains (fig. 1). Fluid inclusions in quartz and barite from the Mayfield prospect have homogenization temperatures (fig. 37) similar to those from silver-copper-lead veins. Because they expand when heated, inclusions in barite have higher homogenization temperatures than do those in quartz. Freezing-point depressions measured on these inclusions are less than 1°C and indicate salinities of less than 2 weight percent equivalent NaCl, unlike the saline fluids that transported and deposited metals in the silver-copper-lead deposits

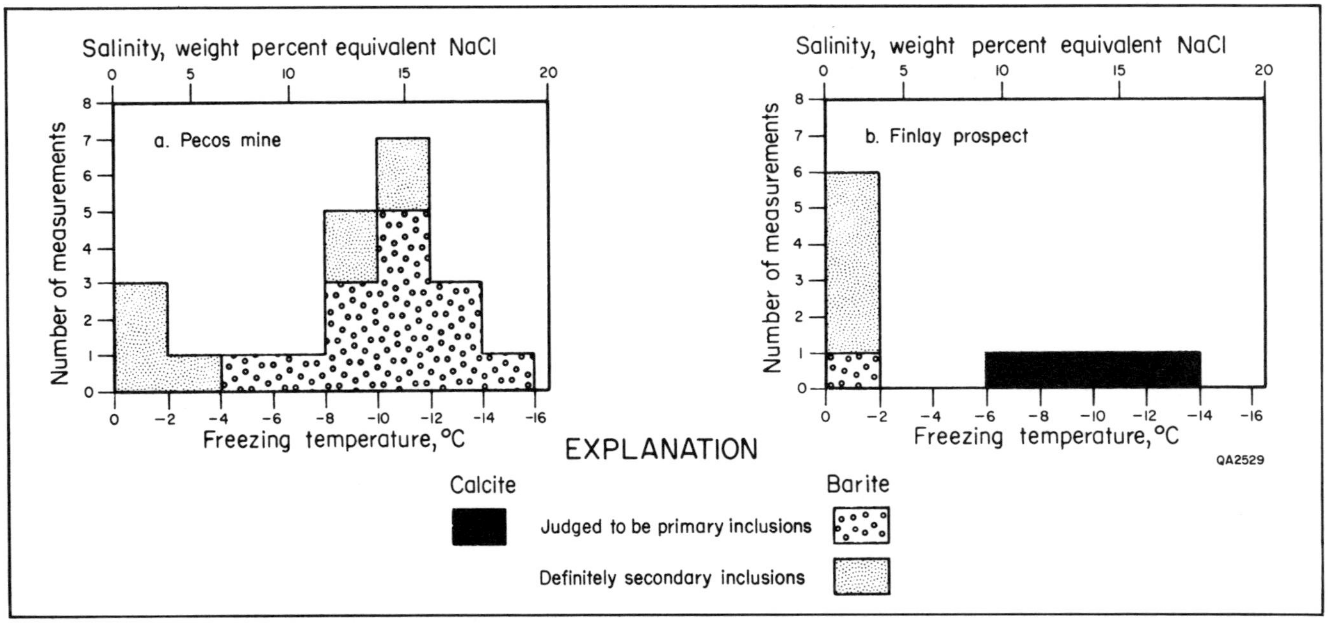

FIGURE 36. Freezing temperatures of fluid inclusions from (a) the Pecos mine in the Precambrian Hazel Formation and (b) the Finlay prospect in the Cretaceous Yucca Formation. Ore-forming fluids at the Pecos mine ranged in salinity from 9 to 19 weight percent equivalent NaCl.

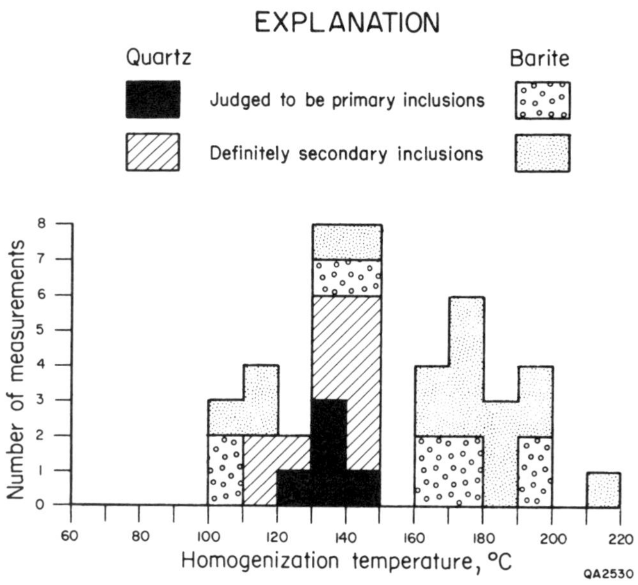

FIGURE 37. Homogenization temperatures of fluid inclusions from the Mayfield manganese prospect, a vein occurrence along a Basin and Range fault. Temperatures similar to those observed at silver-copper-lead deposits in red beds support the hypothesis that the ores in red beds also formed during Basin and Range extension.

of Texas and the galena-fluorite-barite deposits of New Mexico. The moderately high homogenization temperatures, 120°C to 150°C, of primary inclusions in quartz attest to the circulation of geothermal waters along Basin and Range faults.

In the Hansonburg district of New Mexico, galena-barite-fluorite ores were deposited as the result of simple cooling (Putnam and others, 1983). Mineralization occurred in the temperature range of approximately 130°C to 210°C. Although salinities varied from 10 to 18 weight percent equivalent NaCl, Putnam and others (1983) preferred a one-solution model of ore deposition rather than a model involving mixing two fluids.

Carbon and Sulfur Isotope Studies

Carbon isotope data (table 14) provided by Texasgulf Minerals Exploration Company (G. P. Eager, written communication, 1983) are consistent with a moderate-temperature hydrothermal origin of vein mineralization. The low negative $\delta^{13}C$ value for gangue calcite from the Eureka prospect, however, cannot be used alone to define the carbon source (Ohmoto and Rye, 1979). Sedimentary carbonates from the Allamoore and Hazel Formations (table 14) have $\delta^{13}C$ values typical of marine carbonates. Because dissolution and decarbonation reactions tend to yield either isotopically similar or higher $\delta^{13}C$ values (Ohmoto and Rye, 1979), these sedimentary carbonates are not likely sources of the carbon in the mineral deposit.

Sulfur isotope data on barite (table 15), obtained by Kyle (in preparation), are insufficient for determining a source of the sulfur. Additional sulfur isotope studies involving sulfides as well as sulfates are needed to draw concrete conclusions. Given that vein barites from three different areas (the Precambrian Hazel Formation, Cretaceous Yucca Formation, and Cretaceous Loma Plata Limestone) have distinctly different $\delta^{34}S$ values but probably similar formation temperatures, more than one source of sulfur may be involved in these deposits.

TABLE 14. Carbon isotope data from the Eureka prospect and Precambrian carbonates.[1]

Locality/Sample No.	$\delta^{13}C_{PDB}$, ‰
Calcite breccia, Hazel Formation, Eureka prospect, J82-35	-5.4
Stromatolitic limestone, Hazel Formation, Blackshaft mine, J82-15	+0.1
Dolomite, Allamoore Formation, near Dallas prospect, J82-46	+3.4

[1]G. P. Eager (written communication, 1983)

TABLE 15. Sulfur isotope data on vein barite, Van Horn area.[1]

Locality/Sample No.	$\delta^{34}S$, ‰
Precambrian Hazel Formation	
Hazel mine/CDH-1180	14.9
Pecos mine/J82-26	16.0
Cretaceous Yucca Formation	
Finlay prospect/J81-110	6.2
Road prospect/J81-109	6.5
Cretaceous Loma Plata Limestone	
Mayfield manganese prospect/ J81-104	20.5
J83-21	26.4

[1]Kyle (in preparation)

Review of Observations and Conclusions

Data analyzed thus far support the hypothesis that the silver-copper-lead deposits in red-bed sequences near Van Horn formed from moderate-temperature hydrothermal fluids that moved upward along Basin and Range fractures and precipitated minerals in response to mixing with shallow ground water or, less likely, by a drop in temperature. Several features of the deposits in red beds indicate that they did not form as a result of middle Tertiary magmatism, the hypothesis suggested by King and Flawn (1953) and Wallace (1972). (1) The deposits are distant from centers of Tertiary igneous activity. No Tertiary igneous rocks crop out within 10 mi (16 km) of the most productive ores in Precambrian red beds. (2) Limestones, common hosts for igneous-related ores, are generally unmineralized above and below the mineralized red beds. (3) Extensive zones of sericitic, argillic, or propylitic alteration, typical of igneous-hydrothermal veins, are missing. (4) Quartz, the most abundant gangue mineral in Trans-Pecos igneous-hydrothermal veins, is scarce in the veins in red beds. Fluorite, common in igneous-hydrothermal veins in the region, is absent in the red-bed ores. Barite, a mineral typically formed at relatively low temperature (Holland and Malinin, 1979), is the most abundant gangue mineral. (5) Homogenization temperatures of fluid inclusions in the ores suggest formation temperatures in the range of 120°C to 170°C. These values are generally lower than those typical of copper-lead-zinc-bearing igneous-hydrothermal veins (Spooner, 1981), but are appropriate for fluids moving upward along Basin and Range faults. (6) The geochemical suite of elements dominated by silver, copper, and lead is compatible with this relatively low temperature origin (Rose, 1976). The lack of gold is also consistent with relatively low temperature ore deposition. (7) Structural evidence, including orientations of veins, relative ages of fractures, and relations to major tectonic events in the region, suggests that the most likely time of mineral deposition was during late Basin and Range extension, no more than 10 m.y.a. and at least 18 m.y. after silicic volcanism in the region had ceased.

The wide salinity range of barite deposition observed in Precambrian rocks at the Pecos mine and of calcite deposition in Cretaceous rocks at the Finlay prospect (fig. 36) implies that simple cooling of one fluid is a less likely mechanism for ore deposition than is mixing of a hot, upwelling, saline water with shallow ground water. Because the fluids are all undersaturated with respect to halite and because sodium apparently was not lost to the wall rocks during alteration and ore deposition, simple cooling of ore fluid would not be likely to yield the noted variability in salinity. Mixing hot, upwelling, saline water with cooler, oxidized, fresher, shallow ground water could cause precipitation by decomplexing, lowering temperature, or, for barite, oxidation.

Unanswered Questions

Several questions regarding the origin of the ores in red beds remain unanswered. What caused the salinity of the hydrothermal fluids? Evaporites occur in Permian, Mesozoic, and Cenozoic rocks in the region and may exist in the subsurface Allamoore Formation below the Hazel red beds. Did these evaporites yield sodium chloride, or are sodium chloride waters the normal geothermal waters in the region? Henry (1979) suggested that present-day geothermal waters of the region are saline because of dissolution of evaporites at depth. Evaporites in the Mesozoic section of the Chihuahua trough are potential sources of sulfur for the mineralizing fluids in Cretaceous rocks, and evaporites of Precambrian, Permian, or Cenozoic age are possible sources for the ores in the Hazel Formation. The differences in source may account for the observed isotopic differences (table 15); additional stable isotope studies may help to reveal the origin of the sodium chloride and sulfur.

What is the origin of the silver, copper, and other metals in the deposits? Although no unambiguous answer can be provided, several possibilities exist. In the veins in the Hazel Formation, metals from stratigraphically lower Blackshaft-type stratiform deposits (table B4 in the appendix) may have been remobilized by the ore-forming solutions. Geochemical segregation of silver and copper, perhaps resulting from leaching by somewhat oxidizing solutions (Rose, 1976), may explain the difference in copper-to-silver ratios between the Blackshaft-type and Hazel-type deposits (table 3).

Local concentrations of metals in reduced parts of the red-bed sequences are common. Small deposits of red-bed copper protore could have been leached and the metals could have been reconcentrated in the vein deposits. Somewhat anomalous concentrations of copper occur in reduced, greenish-gray Powwow Member sandstone 5 mi (8 km) northeast of the Plata Verde mine (Price, 1982). Clark and de la Fuente (1978) noted stratiform copper-barite deposits in Lower Cretaceous strata in Chihuahua, Mexico. These deposits are in the Las Vigas Formation, the correlative of the Yucca Formation in Texas (Campbell, 1980).

King and Adkins (1946) mentioned that at Mina Las Vigas, the type locality of Las Vigas Formation and the largest of the copper deposits, the nearly vertically dipping rocks contain nodules and "veins" of copper ore. Weed (1902) contended that the "veins" are actually strata-bound layers of disseminated copper minerals in nearly vertical, permeable sandstones. It is unknown how much of the apparent strata-bound copper ore is in fact vein controlled. Descriptions by Weed (1902) of strata-bound Las Vigas ore indicate that the ore is similar to disseminated chalcocite in sandstone at the Rossman prospect in the Yucca Formation (see section on Wall-Rock Alteration, p. 37). Silver and lead are apparently rare in most of the Chihuahuan localities, but barite is common (F. E. de la Fuente, personal communication, 1983). The copper-to-silver ratio in a composite of 18 samples reported by Weed (1902) was 1,050 to 1. Although copper in the vein deposits in the Yucca Formation may have been derived from preexisting strata-bound occurrences, there is no evidence of a similar source of silver or lead.

An alternative source of metals is indigenous detritus in the red beds. Copper and silver deposits in the Allamoore Formation and perhaps in the Carrizo Mountain Group could have provided sedimentary detritus to the Hazel Formation. Additional sources of metals could be fragments of igneous rocks from the Allamoore Formation. Small copper occurrences in the Carrizo Mountain Group near the Plata Verde mine (Price and others, 1983b) may have contributed sediment to the Powwow red beds. Detrital material rich in silver, copper, lead, and related metals (for example, sulfide minerals, mafic igneous minerals, and feldspars) readily weathers chemically. Consequently, if the source of the metals had been the red beds themselves, the least productive deposits would most likely be in the sandstones containing the lowest component of detrital igneous material, those in the Cretaceous Yucca Formation.

Another potential source of metals is Tertiary tuffaceous material that was deposited on top of the red beds and then eroded. This is a less likely source, because chloride-rich brines capable of complexing the metals (as observed in fluid-inclusion freezing studies) probably rose along the fractures and did not percolate downward through the Tertiary rocks.

Lead isotope data, which may help to pinpoint the origin of metals in the silver-copper-lead deposits in Trans-Pecos Texas, are lacking. At present, the question of source, which has practical applications for exploration, remains unanswered.

EXPLORATION POTENTIAL

The hypothesis about the origin of the known silver-copper-lead deposits in red-bed sequences can be used as a guide for exploration near existing mines and elsewhere. Exploration need not be confined to areas of igneous activity because structural control of mineralization was probably Basin and Range faulting, fracturing, and extension of preexisting fractures. Areas favorable for exploration thus include those that underwent prolonged Basin and Range extension and that contain thick, permeable red beds. Obvious regions for search include the three formations that host silver-copper-lead deposits in Texas plus the Precambrian Van Horn Sandstone in Texas, Permian red beds in New Mexico, and Mesozoic clastic sequences in Chihuahua and Coahuila, Mexico.

Geochemical exploration methods may be fruitful because the mineralogy and geochemistry of the ores may yield large dispersion halos. Although primary dispersion halos around the vein deposits are generally small, secondary halos detectable by stream-sediment analyses or other techniques are expected to be large. Barite, the common gangue mineral in the veins, tends to survive mechanical and chemical weathering well in streams near these deposits, and routine analysis for barium and sulfur is advisable. If barite is a good indicator of potentially exploitable silver veins, then scattered occurrences of vein barite should be reexamined. Known vein barite in the Texas part of the Basin and Range province (Price and others, 1983b) includes the Mayfield manganese prospect and three other areas near

Van Horn: (1) veins in Ordovician limestone and Precambrian or Permian sandstone northwest of Van Horn, (2) barite and cerussite in Precambrian and Permian rocks near Eagle Flat west of Van Horn, and (3) barite in the Indio Mountains 3 mi (5 km) west of the silver-copper-lead deposits.

Exploration near known deposits may also be successful, because mining to date has only exploited outcropping orebodies. The vertical extent of ore at the Hazel mine has not been established; stratiform orebodies similar to the Blackshaft-type deposits may lie at depth below the outcropping Hazel, Mohawk, Pecos, and other veins. Exploration at the Plata Verde mine has literally only scratched the surface with backhoe trenches and shallow drill holes. Because some of the secondary ores at the mine are conformable with bedding (Price, 1982), it is reasonable to assume some continuity following the dipping strata. If ore formation involves upward movement of hydrothermal fluids at the Plata Verde mine, then the numerous faults nearby should be explored for high-grade vein mineralization.

ACKNOWLEDGMENTS

Numerous Trans-Pecos mine owners and landowners and their employees provided access to properties. Tye Beams, Dub Garren, Red Pascal, Glen Pepper, Grant Roane, Don Shoemake, Don Switzer, and Phil Beckley were especially helpful.

Chemical analyses were performed by Steven W. Tweedy, Dorothy L. Gower, Paul N. Williams, Josefina R. Calvo, and Nam P. Bui, under the supervision of Clara L. Ho at the Mineral Studies Laboratory of the Bureau of Economic Geology. Long-Cheng Liang assisted with electron microprobe analyses. Phil Eager provided carbon isotope data in conjunction with his regional study in the Delaware Basin. J. Richard Kyle, who is studying barite occurrences in Texas, provided sulfur isotope data.

L. F. Brown, Jr., Jules R. DuBar, Martin P. A. Jackson, J. Richard Kyle, Noel Tyler, and W. A. White critically reviewed this report, which was edited by Amanda R. Masterson. Arthur G. Goldstein provided helpful instructions for collecting structural data and carefully reviewed the manuscript. The manuscript was typed by Dorothy C. Johnson and typeset by Phyllis J. Hopkins, under the direction of Lucille C. Harrell. Figures were drafted by John T. Ames, Mark T. Bentley, Thomas R. Byrd, and Richard M. Platt under the supervision of Richard L. Dillon. James A. Morgan did the text illustration photographic work. The publication was designed by Margaret L. Evans.

Research for this project was funded by the U.S. Department of the Interior, originally through the Office of Surface Mining and later through the U.S. Bureau of Mines, under grant number G114037 to the Texas Mining and Mineral Resources Research Institute.

REFERENCES

Adams, G. B., 1953, Stratigraphy of southern Indio Mountains, Hudspeth County, Texas: University of Texas, Austin, Master's thesis, 64 p.

Albritton, C. C., Jr., and Smith, J. F., Jr., 1965, Geology of the Sierra Blanca area, Hudspeth County, Texas: U.S. Geological Survey Professional Paper 479, 131 p.

Allday, Edwin, 1953, Structure of southern Indio Mountains, Hudspeth County, Texas: University of Texas, Austin, Master's thesis, 31 p.

Barnes, V. E., 1968, Van Horn - El Paso sheet: The University of Texas at Austin, Bureau of Economic Geology, Geologic Atlas of Texas, scale 1:250,000.

_____ 1979, Marfa sheet: The University of Texas at Austin, Bureau of Economic Geology, Geologic Atlas of Texas, scale 1:250,000.

Beane, R. E., 1974, Barite-fluorite-galena deposits in south-central New Mexico: a product of shallow intrusions, ground water, and epicontinental sediments (abs.): Geological Society of America Abstracts with Programs, v. 6, no. 7, p. 646-647.

Berge, T. B. S., 1981, Structural evolution of the Malone Mountains, Hudspeth County, Texas: The University of Texas at Austin, Master's thesis, 95 p.

Berger, B. R., and Eimon, P. I., 1983, Conceptual models of epithermal precious metal deposits, *in* Shanks, W. C., III, ed., Cameron volume on unconventional mineral deposits: New York, Society of Mining Engineers of AIME, p. 191-205.

Bostwick, D. L., 1953, Structural geology of the northern Indio Mountains, Hudspeth County, Texas: University of Texas, Austin, Master's thesis, 56 p.
Botinelly, Theodore, 1976, A review of the minerals of the alunite-jarosite, beudantite, and plumbogummite groups: U.S. Geological Survey Journal of Research, v. 4, no. 2, p. 213-216.
Campbell, D. H., 1980, The Yucca Formation—Early Cretaceous continental and transitional environments, southern Quitman Mountains, Hudspeth County, Texas, *in* Dickerson, P. W., and Hoffer, J. M., eds., Trans-Pecos region, southwestern New Mexico and West Texas: New Mexico Geological Society 31st Field Conference Guidebook, p. 159-168.
Cepeda, J. C., and Henry, C. D., 1983, Oligocene volcanism and multiple caldera formation in the Chinati Mountains, Presidio County, Texas: The University of Texas at Austin, Bureau of Economic Geology Report of Investigations No. 135, 32 p.
Chapin, C. E., 1979, Evolution of the Rio Grande rift—a summary, *in* Riecker, R. E., ed., Rio Grande rift: tectonics and magmatism: American Geophysical Union, p. 1-5.
Chapin, C. E., and Cather, S. M., 1981, Eocene tectonics and sedimentation in the Colorado Plateau - Rocky Mountains area, *in* Dickinson, W. R., and Payne, W. D., eds., Relations of tectonics to ore deposits in the southern Cordillera: Arizona Geological Society Digest, v. 14, p. 199-213.
Chuchla, R. J., 1981, Reconnaissance geology of the Sierra Rica area, Chihuahua, Mexico: The University of Texas at Austin, Master's thesis, 199 p.
Clark, K. F., and de la Fuente L., F. E., 1978, Distribution of mineralization in time and space in Chihuahua, Mexico: Mineralium Deposita, v. 13, p. 27-49.
Condie, K. C., 1982, Plate-tectonics model for Proterozoic continental accretion in the southwestern United States: Geology, v. 10, p. 37-42.
Dasch, E. J., Jr., 1959, Dike swarm of northern Rim Rock Country, Trans-Pecos Texas: University of Texas, Austin, Master's thesis, 62 p.
Dasch, E. J., Armstrong, R. L., and Clabaugh, S. E., 1969, Age of Rim Rock dike swarm, Trans-Pecos Texas: Geological Society of America Bulletin, v. 80, p. 1819-1824.
Dayton, S. H., 1983, Asarco's Troy mine, how rolling the dice in a new district can add to earnings: Engineering and Mining Journal, v. 184, no. 9, p. 40-46.
Decamp, D. W., 1981, Structural geology of Mesa de Anguila, Big Bend National Park, Trans-Pecos Texas: The University of Texas at Austin, Master's thesis, 185 p.
Denison, R. E., 1980, Pre-Bliss (p€) rocks in the Van Horn region, Trans-Pecos Texas, *in* Dickerson, P. W., and Hoffer, J. M., eds., Trans-Pecos region, southwestern New Mexico and West Texas: New Mexico Geological Society 31st Field Conference Guidebook, p. 155-158.
Denison, R. E., and Hetherington, E. A., Jr., 1969, Basement rocks in far West Texas and south-central New Mexico, *in* Kottlowski, F. E., and LeMone, D. V., eds., Border stratigraphy symposium: New Mexico Bureau of Mines and Mineral Resources Circular 104, p. 1-14.
Dickinson, W. R., 1981, Plate tectonic evolution of the southern Cordillera, *in* Dickinson, W. R., and Payne, W. D., eds., Relations of tectonics to ore deposits in the southern Cordillera: Arizona Geological Society Digest, v. 14, p. 113-135.
Dumas, D. B., 1981, Seismicity of West Texas: The University of Texas at Dallas, Ph.D. dissertation, 94 p.
Dumas, D. B., Dorman, H. J., and Latham, G. V., 1980, A reevaluation of the August 16, 1931, Texas earthquake: Bulletin of the Seismological Society of America, v. 70, no. 4, p. 1171-1180.
Dunham, K. C., 1935, The geology of the Organ Mountains: New Mexico School of Mines, State Bureau of Mines and Mineral Resources Bulletin No. 11, 272 p.
Evans, T. J., 1975, Gold and silver in Texas: The University of Texas at Austin, Bureau of Economic Geology Mineral Resource Circular No. 56, 36 p.
Flawn, P. T., 1950, Squaw Springs area: University of Texas, Austin, Bureau of Economic Geology, unpublished notes on Hudspeth County, 1 p.
_____ 1952, The Hazel copper-silver mine, Culberson County, Texas: University of Texas, Austin, Bureau of Economic Geology Report of Investigations No. 16, 21 p.
Galley, J. E., 1958, Oil and geology in the Permian Basin of Texas and New Mexico, *in* Weeks, L. G., ed., Habitat of oil: American Association of Petroleum Geologists, p. 395-446.
Gillerman, Elliot, 1953, Fluorspar deposits of the Eagle Mountains, Trans-Pecos Texas: U.S. Geological Survey Bulletin 987, 98 p.
Goldstein, A. G., 1981, Comment on "Plate tectonics of the Ancestral Rocky Mountains": Geology, v. 9 (supp.), p. 387-388.
Golombek, Matthew, 1982, Extension in the Rio Grande rift and Basin and Range: active or passive? (abs.): Geological Society of America Abstracts with Programs, v. 14, no. 7, p. 498.
_____ 1983, Rate of extension across the Española Basin, Rio Grande rift, New Mexico (abs.): Geological Society of America Abstracts with Programs, v. 15, no. 5, p. 321.
Gregory, J. L., 1981, Volcanic stratigraphy and K-Ar ages of the Manuel Benavides area, northeastern Chihuahua, Mexico, and corre-

lations with the Trans-Pecos Texas volcanic province: The University of Texas at Austin, Master's thesis, 78 p.
Harbour, R. L., 1972, Geology of the northern Franklin Mountains, Texas and New Mexico: U.S. Geological Survey Bulletin 1298, 129 p.
Hay-Roe, Hugh, 1957, Geology of Wylie Mountains and vicinity, Culberson and Jeff Davis Counties, Texas: University of Texas, Austin, Bureau of Economic Geology Geologic Quadrangle Map No. 21, scale 1:63,360.
Heidrick, T. L., and Titley, S. R., 1982, Fracture and dike patterns in Laramide plutons and their structural and tectonic implications: American Southwest, *in* Titley, S. R., ed., Advances in geology of porphyry copper deposits, southwestern North America: Tucson, The University of Arizona Press, p. 73-91.
Henderson, C. W., and Mote, R. H., 1945, Gold, silver, copper, lead, and zinc in Texas: U.S. Bureau of Mines Minerals Yearbook 1943, p. 457-460.
Henry, C. D., 1979, Geologic setting and geochemistry of thermal water and geothermal assessment, Trans-Pecos Texas, with tectonic map of the Rio Grande area, Trans-Pecos Texas and adjacent Mexico: The University of Texas at Austin, Bureau of Economic Geology Report of Investigations No. 96, 48 p.
Henry, C. D., and McDowell, F. W., 1982, Timing, distribution, and estimates of volumes of silicic volcanism in Trans-Pecos Texas (abs.): Geological Society of America Abstracts with Programs, v. 14, no. 3, p. 113.
Henry, C. D., and Price, J. G., 1984, Variations in caldera development in the Tertiary volcanic field of Trans-Pecos Texas: Journal of Geophysical Research, v. 89, no. B 10, p. 8765-8786.
Henry, C. D., Price, J. G., and McDowell, F. W., 1983, Presence of the Rio Grande rift in West Texas and Chihuahua, *in* Clark, K. F., and Goodell, P. C., eds., Geology and mineral resources of north-central Chihuahua: El Paso Geological Society Guidebook for the 1983 Field Conference, p. 108-118.
Holland, H. D., and Malinin, S. D., 1979, The solubility and occurrence of non-ore minerals, *in* Barnes, H. L., ed., Geochemistry of hydrothermal ore deposits (2d ed.): New York, John Wiley, p. 461-508.
Immitt, J. P., 1981, Skarn and epithermal vein mineralization in the San Carlos caldera region, northeastern Chihuahua, Mexico: The University of Texas at Austin, Master's thesis, 110 p.
Ingersoll, R. V., 1982, Triple-junction instability as a cause for late Cenozoic extension and fragmentation of the western United States: Geology, v. 10, p. 621-624.
Jones, B. R., and Reaser, D. F., 1970, Geology of southern Quitman Mountains, Hudspeth County, Texas: The University of Texas at Austin, Bureau of Economic Geology Geologic Quadrangle Map No. 39, scale 1:48,000, 24-p. text.
King, P. B., 1937, Geology of the Marathon region, Texas: U.S. Geological Survey Professional Paper 187, 147 p.
_____ 1965, Geology of the Sierra Diablo region, Texas: U.S. Geological Survey Professional Paper 480, 185 p.
_____ 1980, Geology of the eastern part of the Marathon Basin, Texas: U.S. Geological Survey Professional Paper 1157, 40 p.
King, P. B., and Flawn, P. T., 1953, Geology and mineral deposits of Pre-Cambrian rocks of the Van Horn area, Texas: Austin, University of Texas Publication 5301, 218 p.
King, R. E., and Adkins, W. S., 1946, Geology of a part of the lower Conchos valley, Chihuahua, Mexico: Geological Society of America Bulletin, v. 57, p. 275-294.
Kluth, C. F., and Coney, P. J., 1981, Plate tectonics of the Ancestral Rocky Mountains: Geology, v. 9, p. 10-15.
Kyle, J. R., in preparation, The barite industry and resources of Texas: The University of Texas at Austin, Bureau of Economic Geology Mineral Resource Circular.
Lange, I. M., and Sherry, R. A., 1983, Genesis of the sandstone (Revett) type of copper-silver occurrences in the Belt Supergroup of northwestern Montana and northeastern Idaho: Geology, v. 11, p. 643-646.
Laughlin, A. W., Aldrich, M. J., and Vaniman, D. T., 1983, Tectonic implications of mid-Tertiary dikes in west-central New Mexico: Geology, v. 11, p. 45-48.
Laughlin, A. W., Kress, V. C., and Aldrich, M. J., 1982, K-Ar ages of dike rocks, Big Bend National Park, Texas: Isochron West, no. 35, p. 17-18.
Laux, J. P., 1969, Mineralization associated with the Quitman Mountains intrusion, Hudspeth County, Texas: The University of Texas at Austin, Master's thesis, 86 p.
Lemmlein, G. G., and Klevtsov, P. V., 1961, Relations among the principal thermodynamic parameters in a part of the system H_2O-NaCl: Geochemistry, no. 2, p. 148-158.
LeMone, D. V., Kondelin, R., McEvers, L., Madden, H. D., Roark, R. C., Simpson, R. D., and Pickens, C. A., 1983, Paleozoic and Early Cretaceous strata of the El Paso borderland, *in* Meader-Roberts, S. J., ed., Geology of the Sierra Diablo and southern Hueco Mountains, West Texas: Society of Economic Paleontologists and Mineralogists, Permian Basin Section Publication 83-22, p. 122-130.
Lipman, P. W., 1981, Volcano-tectonic setting of Tertiary ore deposits, Southern Rocky Mountains, *in* Dickinson, W. R., and Payne, W. D.,

eds., Relations of tectonics to ore deposits in the southern Cordillera: Arizona Geological Society Digest, v. 14, p. 199-213.

Lucchita, Ivo, and Suneson, Neil, 1983, Mid- and late-Cenozoic extensional tectonism near the Colorado Plateau boundary in west-central Arizona (abs.): Geological Society of America Abstracts with Programs, v. 15, no. 5, p. 405.

Maxwell, R. A., Lonsdale, J. T., Hazzard, R. T., and Wilson, J. A., 1967, Geology of Big Bend National Park, Brewster County, Texas: The University of Texas at Austin, Bureau of Economic Geology Publication 6711, 320 p.

McAnulty, W. N., Sr., 1972, Mineral deposits in the West Chinati stock, Chinati Mountains, Presidio County, Texas: The University of Texas at Austin, Bureau of Economic Geology Geological Circular 72-1, 13 p.

McDowell, F. W., 1979, Potassium-argon dating in the Trans-Pecos Texas volcanic field, *in* Walton, A. W., and Henry, C. D., eds., Cenozoic geology of the Trans-Pecos volcanic field of Texas: The University of Texas at Austin, Bureau of Economic Geology Guidebook 19, p. 10-18.

McGowen, J. H., and Groat, C. G., 1971, Van Horn Sandstone, West Texas: an alluvial fan model for mineral exploration: The University of Texas at Austin, Bureau of Economic Geology Report of Investigations No. 72, 57 p.

McKnight, J. F., 1970, Geology of Bofecillos Mountains area, Trans-Pecos Texas: The University of Texas at Austin, Bureau of Economic Geology Geologic Quadrangle Map No. 37, scale 1:48,000, 36-p. text.

McMillan, W. D., 1949, Investigation of Montezuma and Chinati zinc-lead deposits, Shafter district, Presidio County, Texas: U.S. Bureau of Mines Report of Investigations 4506, 13 p.

Morgan, Paul, and Seager, W. R., 1983, Thermal, mechanical, and tectonic evolution of the southern Rio Grande rift (abs.): Geological Society of America Abstracts with Programs, v. 15, no. 5, p. 320.

Moustafa, A. R., 1983, Analysis of Laramide and younger deformation of a segment of the Big Bend region, Texas: The University of Texas at Austin, Ph.D. dissertation, 157 p.

Muehlberger, W. R., 1980, Texas lineament revisited, *in* Dickerson, P. W., and Hoffer, J. M., eds., Trans-Pecos region, southwestern New Mexico and West Texas: New Mexico Geological Society 31st Field Conference Guidebook, p. 113-121.

Murry, D. H., 1979, Economic mineral potential of the northern Quitman Mountains, Hudspeth County, Texas: The University of Texas at El Paso, Master's thesis, 156 p.

Ogley, D. S., 1978, Eruptive history of the Pine Canyon caldera, Big Bend National Park, Texas: The University of Texas at Austin, Master's thesis, 124 p.

Ohmoto, Hiroshi, and Rye, R. O., 1979, Isotopes of sulfur and carbon, *in* Barnes, H. L., ed., Geochemistry of hydrothermal ore deposits (2d ed.): New York, John Wiley, p. 509-561.

Parker, D. F., Jr., 1976, Petrology and eruptive history of an Oligocene trachytic shield volcano, near Alpine, Texas: The University of Texas at Austin, Ph.D. dissertation, 183 p.

——— 1979, The Paisano volcano: stratigraphy, age, and petrogenesis, *in* Walton, A. W., and Henry, C. D., eds., Cenozoic geology of the Trans-Pecos volcanic field of Texas: The University of Texas at Austin, Bureau of Economic Geology Guidebook 19, p. 97-101.

Price, J. G., 1982, Geology of the Plata Verde mine, Hudspeth County, Texas: The University of Texas at Austin, Bureau of Economic Geology Mineral Resource Circular No. 70, 34 p.

——— 1983, Silver-copper-lead deposits in red-bed sequences of Trans-Pecos Texas: Tertiary mineralization in Precambrian, Permian, and Cretaceous sandstones: Society of Mining Engineers of AIME Preprint 83-3, 11 p.

Price, J. G., Henry, C. D., Posey, J. S., and Standen, A. R., 1983a, Structure, geochemistry, mineralogy, and origin of silver-copper-lead deposits in the Indio Mountains, Trans-Pecos Texas (abs.): Geological Society of America Abstracts with Programs, v. 15, no. 1, p. 14.

Price, J. G., Henry, C. D., and Standen, A. R., 1983b, Annotated bibliography of mineral deposits in Trans-Pecos Texas: The University of Texas at Austin, Bureau of Economic Geology Mineral Resource Circular No. 73, 108 p.

Price, J. G., and Henry, C. D., 1984, Stress orientations during Oligocene volcanism in Trans-Pecos Texas: timing the transition from Laramide compression to Basin and Range extension: Geology, v. 12, p. 238-241.

Putnam, B. R., III, Norman, D. I., and Smith, R. W., 1983, Mississippi Valley-type lead-fluorite-barite deposits of the Hansonburg mining district, *in* Chapin, C. E., and Callender, J. F., eds., Socorro Region II: New Mexico Geological Society 34th Annual Field Conference Guidebook, p. 253-259.

Rehrig, W. A., and Heidrick, T. L., 1976, Regional tectonic stress during the Laramide and late Tertiary intrusive periods, Basin and Range province, Arizona: Arizona Geological Society Digest, v. 10, p. 205-228.

Reid, J. C., 1974, Hazel Formation, Culberson and Hudspeth Counties, Texas: The University of Texas at Austin, Master's thesis, 88 p.

Richardson, G. B., 1914, Description of the Van Horn quadrangle: U.S. Geological Survey Geologic Atlas (folio no. 194), 9 p.

Roedder, Edwin, 1979, Fluid inclusions as samples of ore fluids, *in* Barnes, H. L., ed., Geochemistry of hydrothermal ore deposits (2d ed.): New York, John Wiley, p. 684-737.

Rose, A. W., 1976, The effect of cuprous chloride complexes in the origin of red-bed copper and related deposits: Economic Geology, v. 71, p. 1036-1048.

Rose, A. W., Hawkes, H. E., and Webb, J. S., 1979, Geochemistry in mineral exploration (2d ed.): London, Academic Press, 657 p.

Ross, C. P., 1943, Geology and ore deposits of the Shafter mining district, Presidio County, Texas: U.S. Geological Survey Bulletin 928-B, p. 45-125.

Rudnick, R. L., 1983, Geochemistry and tectonic affinities of a Proterozoic bimodal igneous suite, West Texas: Geology, v. 11, p. 352-355.

Sample, R. D., and Gould, E. E., 1945, Geology and ore deposits of the Allamoore - Van Horn copper district, Hudspeth and Culberson Counties, Texas: U.S. Geological Survey Open-File Report, 22 p. (on open file at The University of Texas at Austin, Bureau of Economic Geology).

Scott, R. W., and Kidson, E. J., 1977, Lower Cretaceous depositional systems, West Texas, *in* Bebout, D. G., and Loucks, R. G., eds., Cretaceous carbonates of Texas and Mexico, applications to subsurface exploration: The University of Texas at Austin, Bureau of Economic Geology Report of Investigations No. 89, p. 169-181.

Seager, W. R., 1980, Quaternary fault system in the Tularosa and Hueco Basins, southern New Mexico and West Texas, *in* Dickerson, P. W., and Hoffer, J. M., eds., Trans-Pecos region, southwestern New Mexico and West Texas: New Mexico Geological Society 31st Field Conference Guidebook, p. 131-135.

Sellards, E. H., and Evans, G. L., 1946, Index to Texas mineral resources, *in* Texas mineral resources: Austin, University of Texas Publication 4301, p. 359-383.

Spooner, E. T. C., 1981, Fluid inclusion studies of hydrothermal ore deposits, *in* Hollister, L. S., and Crawford, M. L., eds., Short course in fluid inclusions: applications to petrology: Calgary, Mineralogical Association of Canada, p. 209-240.

Thomann, W. F., 1981, Ignimbrites, trachytes, and sedimentary rocks of the Precambrian Thunderbird Group, Franklin Mountains, El Paso, Texas: Geological Society of America Bulletin, part I, v. 92, p. 94-100.

Twiss, P. C., 1959a, Geology of Van Horn Mountains, Trans-Pecos Texas: University of Texas, Austin, Ph.D. dissertation, 234 p.

_____ 1959b, Geology of Van Horn Mountains, Texas: University of Texas, Austin, Bureau of Economic Geology Geologic Quadrangle Map No. 23, scale 1:48,000.

Ulrich, M. R., and Bodnar, J. R., in press, Systematics of stretching of fluid inclusions II: barite at 1 atm confining pressure: Economic Geology.

Underwood, J. R., Jr., 1962, Geology of Eagle Mountains and vicinity, Trans-Pecos Texas: University of Texas, Austin, Ph.D. dissertation, 560 p.

_____ 1963, Geology of Eagle Mountains and vicinity, Hudspeth County, Texas: University of Texas, Austin, Bureau of Economic Geology Geologic Quadrangle Map No. 26, scale 1:48,000, 32-p. text.

_____ 1980, Geology of the Eagle Mountains, Hudspeth County, Texas, *in* Dickerson, P. W., and Hoffer, J. M., eds., Trans-Pecos region, southwestern New Mexico and West Texas: New Mexico Geological Society 31st Field Conference Guidebook, p. 183-193.

von Streeruwitz, W. H., 1890, Geology of Trans-Pecos Texas: Geological Survey of Texas First Annual Report (1889), p. 224-235.

_____ 1891, Report on the geology and mineral resources of Trans-Pecos Texas: Geological Survey of Texas Second Annual Report (1890), p. 665-713.

_____ 1892a, Report for the second report of progress: Geological Survey of Texas (1891), p. 20-26.

_____ 1892b, Trans-Pecos Texas: Geological Survey of Texas Third Annual Report (1891), p. 383-389.

_____ 1893, Trans-Pecos Texas: Geological Survey of Texas Fourth Annual Report (1892), p. 141-167.

Walenta, Kurt, 1966, Beitraege zur Kentnis seltener Arsenatmineralien unter besonderer Beruecksichtigung von Vorkommen des Schwarzwaldes: Tschermaks Mineralogische und Petrographische Mitteilungen, v. 11, p. 121-164.

Wallace, A. B., 1972, Mineral deposits of the Indio Mountains, Hudspeth County, Texas: The University of Texas at El Paso, Master's thesis, 49 p.

Wallace, A. B., and Shannon, S. S., 1975, Mineral deposits of the central Indio Mountains, Hudspeth County, Texas: Society of Economic Paleontologists and Mineralogists, Permian Basin Section Guidebook Publication 75-15, p. 144-147.

Warren, L. E., 1946, Manganese deposits of Texas, *in* Texas mineral resources: Austin, University of Texas Publication 4301 (1943), p. 249-257.

Wedepohl, K. H., ed., 1969, Handbook of geochemistry, v. 1: Berlin, Springer-Verlag, 442 p.

Weed, W. H., 1902, Notes on certain mines in the states of Chihuahua, Sinaloa, and Sonora, Mexico: American Institute of Mining and Metallurgical Engineers Transactions, v. 32, p. 396-443.
Wilson, J. A., 1971, Vertebrate biostratigraphy of Trans-Pecos Texas and northern Mexico, *in* The geologic framework of the Chihuahua tectonic belt: West Texas Geological Society, p. 157-166.
Yates, R. G., and Thompson, G. A., 1959, Geology and quicksilver deposits of the Terlingua district, Texas: U.S. Geological Survey Professional Paper 312, 114 p.
Zoback, M. L., Anderson, R. E., and Thompson, G. A., 1981, Cainozoic evolution of the state of stress and style of tectonism of the Basin and Range province of the western United States: Philosophical Transactions of the Royal Society of London, v. 300, p. 407-434.
Zoback, M. L., and Thompson, G. A., 1978, Basin and Range rifting in northern Nevada; clues from a mid-Miocene rift and its subsequent offsets: Geology, v. 6, p. 111-116.

APPENDIX

Structural and Chemical Data

Table A1 lists data on minor faults in the Indio Mountains, as discussed in the Regional Geology—Structure and Tectonics section of this report (p. 14). Geochemical data on Precambrian (tables B1 through B4) and Cretaceous (tables B5 through B8) rocks are discussed in the section on Mineralogy and Geochemistry of Ores and Host Rocks (p. 31). Data on Permian rocks, analytical techniques, and comparisons with standard reference materials were presented by Price (1982).

Table A1. Minor fault orientations in the Indio Mountains.

Fault			Slickenlines		Possible displacement[1]			
					Laramide compression		Basin and Range tension	
Trend	Strike	Dip	Direction	Angle	Early σ_1 = NE	Late σ_1 = ENE	Early σ_3 = ENE	Late σ_3 = NW
N	N3°E	47°E	N89°E	44°		r.rl.ds.	n.ll.ds.	
	N1°W	48°W	S81°W	48°		r.rl.ds.	n.ll.ds.	
	N8°E	57°E	N38°E	41°		r.rl.os.	n.ll.os.	
	N9°E	64°E	N30°E	13°	r.rl.ss.			
	N8°E	70°E	N10°E	10°	r.rl.ss.			
	N2°E	34°E	N90°E	34°		r.rl.ds.	n.ll.ds.	
NNE	N30°E	72°E	S50°E	71°				n.rl.ds.[2]
	N22°E	51°E	S40°E	46°				n.rl.ds.
	N26°E	68°E	S10°E	59°				n.rl.ds.
	N15°E	71°E	S75°E	71°		r.ds.	n.ds.	n.ds.
	N14°E	28°E	S70°E	28°				n.rl.ds.
	N14°E	26°E	N67°E	16°		r.rl.os.	n.ll.os.	
NE	N32°E	74°E	S86°E	60°	r.rl.ds.[3]	r.rl.ds.[3]		
	N43°E	64°E	S88°E	54°			n.ll.ds.	
	N40°E	71°E	S37°E	66°				n.rl.ds.[4]
	N59°E	63°S	N82°E	34°			n.ll.os.	
	N59°E	65°N	N56°E	18°		n.rl.ss.		
	N35°E	50°W	N15°E	20°		n.rl.ss.		
	N45°E	72°E	N57°E	7°		r.rl.ss.		
	N52°E	74°N	S59°W	19°		r.rl.ss.		
ENE	N71°E	52°S	S36°E	52°				n.ll.ds.
	N79°E	89°S	S19°E	89°				n.ll.ds.
	N64°E	60°N	N25°W	60°	r.ll.ds.	n.rl.ds.		n.ds.
	N60°E	90°	N60°E	14°	ll.ss.			
	N60°E	73°S	N60°E	7°	n.ll.ss.			
	N68°E	46°N	N68°E	0°	ll.ss.			
	N70°E	73°N	N65°E	22°	r.ll.ss.			
	N77°E	66°N	N63°E	3°	r.ll.ss.			
	N70°E	90°	N70°E	23°	ll.ss.			
	N61°E	33°S	N85°E	11°	n.ll.ss.			
	N74°E	37°N	S74°W	8°	n.ll.ss.			
	N73°E	35°N	N73°E	0°	ll.ss.			
	N71°E	30°N	N54°E	4°	r.ll.ss.			

Table A1. cont.

Fault			Slickenlines		Possible displacement[1]			
					Laramide compression		Basin and Range tension	
Trend	Strike	Dip	Direction	Angle	Early σ_1 = NE	Late σ_1 = ENE	Early σ_3 = ENE	Late σ_3 = NW
E	N80°W	76°N	N10°E	76°	r.ds.			n.ds.
	N84°E	75°N	N22°W	73°				n.ll.ds.
	N89°E	75°N	N25°E	75°			n.rl.ds.	
	N89°W	75°N	N15°E	75°			n.rl.ds.	
	N83°W	86°S	S7°W	86°	r.ds.[5]	r.ds.[5]		
	N82°W	68°S	S6°E	68°				n.ll.ds.[6]
	N80°W	57°S	S43°W	55°			n.rl.ds.[2]	
	N84°E	86°N	N6°W	86°	r.ds.			n.ds.
	N85°W	90°	N85°W	35°	ll.os.	ll.os.	rl.os.	ll.os.
	N85°W	71°N	N59°E	50°	r.ll.os.		n.rl.os.	
	N84°E	80°S	N90°E	2°	n.ll.ss.	n.ll.ss.		
	N85°W	89°S	N90°W	8°	r.ll.ss.	r.ll.ss.		
WNW	N71°W	80°N	N19°E	80°	r.ds.		n.ds.	
	N66°W	70°S	S13°W	70°				n.ll.ds.
	N76°W	53°S	S60°W	36°	r.ll.os.		n.rl.os.	
	N68°W	88°N	N68°W	2°		n.ll.ss.		
	N79°W	85°S	S75°E	11°		n.ll.ss.		
	N74°W	51°N	N74°W	0°		ll.ss.		
	N68°W	41°N	N21°E	41°	r.ds.	r.ds.		n.ll.ds.
	N74°W	16°N	N84°E	13°	r.ll.os.		n.rl.os.	
NW	N34°W	63°E	N61°E	63°			n.rl.ds.[4]	
	N34°W	75°E	S63°E	43°		r.ll.ds.	n.rl.ds.	
	N46°W	83°E	N43°E	83°	r.ds.	r.ds.	n.ds.	
	N50°W	51°S	N65°W	24°		r.ll.os.	n.rl.os.	
	N38°W	59°E	S60°E	15°		r.ll.ss.		
	N45°W	90°	N45°W	6°		ll.ss.		
	N56°W	75°S	N56°W	0°		ll.ss.		
	N45°W	26°S	S45°W	26°	r.ds.		n.ds.	
	N47°W	42°S	S50°W	38°	r.ll.ds.	r.ll.ds.	n.rl.ds.	
	N45°W	10°S	S54°W	10°	r.ll.ds.	r.ll.ds.	n.rl.ds.	
	N53°W	33°N	N83°E	17°		r.ll.os.	n.rl.os.	
	N59°W	28°N	N59°W	0°		ll.ss.		
NNW	N12°W	90°	N78°E	90°		r.ds.	n.ds.	
	N30°W	60°W	N90°W	55°		r.ll.ds.	n.rl.ds.	
	N20°W	28°W	N81°W	21°		r.ll.ds.		n.rl.ds.
	N20°W	23°E	N56°E	23°	r.ll.ds.			
	N25°W	42°W	N50°W	22°				n.rl.os.

[1]Displacement: n. = normal; r. = reverse; rl. = right lateral; ll. = left lateral; ds. = dip slip; os. = oblique slip (rake between 30 and 60 degrees); ss. = strike slip.
[2]Slickenlines suggest normal, right-lateral displacement.
[3]Slickenlines suggest reverse, right-lateral displacement.
[4]Normal, right-lateral motion indicated by displacement of beds.
[5]Slickenlines suggest reverse displacement.
[6]Slickenlines suggest normal, left-lateral displacement.

Table B1. Geochemical data on ores in the Hazel Formation (values in ppm by weight).

	Mohawk mine		Hazel mine		Pecos mine		
	wall-rock limonitic sandstone	**barite-calcite-malachite vein**	**barite-malachite vein**	**djurleite-bornite-pyrite-marcasite-covellite-barite vein**	**tennantite-chalcopyrite-galena-sphalerite-covellite-marcasite-pyrite-barite-calcite ore from dump**		
Element	**Sample No. 81-163**	**Sample No. 81-167**	**Sample No. 81-177**	**Sample No. 81-179**	**Sample No. 81-123**	**Sample No. 82-27**	**Sample No. 82-28**
Li	88	43	67	10	<1	<1	<1
Be	3.2	2.4	1.5	1.6	0.6	0.5	0.4
Na	5.7×10^2	6.6×10^3	1.9×10^2	1.1×10^2	59	49	71
Mg	5.7×10^3	3.3×10^4	6.4×10^3	4.3×10^2	3.3×10^3	44	4.5×10^3
Al	5.5×10^4	5.0×10^4	3.4×10^4	4.5×10^3	5.0×10^2	1.4×10^2	5.7×10^2
P	6.6×10^2	6.2×10^2	4.3×10^2	4.6×10^3	1.5×10^3	8.0×10^2	3.7×10^2
K	2.7×10^4	1.9×10^4	1.6×10^4	2.2×10^3	78	1.7×10^2	2.4×10^2
Ca	3.4×10^4	4.3×10^4	1.1×10^4	4.3×10^3	9.0×10^3	1.4×10^2	7.3×10^3
Ti	4.3×10^3	3.6×10^3	2.6×10^3	3.4×10^2	24	16	31
V	1.2×10^2	93	50	15	22	43	67
Cr	6×10^1	4×10^1	3×10^1	4	<1	<1	<1
Mn	1.7×10^3	1.3×10^3	6.6×10^2	30	2.9×10^2	4	3.0×10^2
Fe	3.4×10^4	2.0×10^4	7.4×10^3	4.3×10^4	2.2×10^4	2.6×10^4	1.9×10^4
Ni	20	20	10	40	7	5	7
Cu	1.2×10^4	1.1×10^4	8.0×10^3	2.3×10^5	7.2×10^4	4.5×10^4	1.7×10^4
Zn	6.5×10^2	3.4×10^2	1.9×10^2	1.9×10^3	1.9×10^5	3.3×10^3	6.4×10^4
As	1.0×10^3	<5	6	1.0×10^3	2.4×10^4	1.2×10^4	9.0×10^2
Sr	1.4×10^2	1.1×10^2	2.0×10^2	2.9×10^2	4	6	10
Zr	1.0×10^2	9×10^1	7×10^1	10	<1	<1	3
Mo	10	<2.5	<2.5	40	70	1.3×10^1	64
Ag	1.65×10^3	2.2	10	1.50×10^3	71	78	43
Cd	10	<1	<1	4	2.8×10^3	38	460
Au	<0.01	<0.01	<0.01	<0.01	<0.01	<0.01	<0.01
Pb	6×10^1	4×10^1	<10	4×10^1	3×10^3	3×10^3	5×10^3
Th	<5	<5	<5	<5	<5	<5	<5
U	9	<5	<5	<5	<5	<5	<5

Table B2. Geochemical data on weakly mineralized sandstone and siltstone wall rock of the Hazel Formation (values in ppm by weight).

	Mohawk mine	Marvin-Judson prospect		Hazel mine		Pecos mine		
	red ss,* 1 ft from vein	limonitic ss within 6 ft of vein		red ss from dump	limonitic sts** within 5 ft of vein	red ss from dump	gray ss with pyrite-marcasite veinlets, from dump	
Element	Sample No.81-164	Sample No. 82-10	Sample No. 82-11	Sample No. 81-178	Sample No. 82-6	Sample No. 81-181	Sample No. 81-180	Sample No. 82-31
Li	52	30	7.8	19	20	21	32	39
Be	2.3	1.5	1.6	3.4	2.6	2.6	2.0	1.4
Na	2.4×10^4	6.1×10^2	2.8×10^2	3.7×10^2	7.3×10^2	2.8×10^4	2.1×10^4	4.0×10^2
Mg	7.0×10^3	5.9×10^3	4.8×10^4	1.7×10^4	2.0×10^4	9.9×10^3	1.1×10^4	3.0×10^4
Al	6.4×10^4	6.7×10^4	2.0×10^4	6.6×10^4	7.1×10^4	6.6×10^4	6.3×10^4	4.5×10^4
P	5.6×10^2	6.0×10^2	2.1×10^2	6.4×10^2	6.3×10^2	60	5.0×10^2	5.0×10^2
K	2.2×10^4	2.4×10^4	7.4×10^3	2.6×10^4	2.9×10^4	2.3×10^4	2.1×10^4	1.5×10^4
Ca	3.6×10^4	1.5×10^4	$>4 \times 10^4$	2.8×10^4	3.0×10^4	1.7×10^4	1.8×10^4	$>4 \times 10^4$
Ti	5.1×10^3	5.0×10^3	1.5×10^3	5.0×10^3	5.7×10^3	4.3×10^3	3.0×10^3	2.9×10^3
V	1.1×10^2	76	67	1.0×10^2	99	89	68	79
Cr	8×10^1	5×10^1	2×10^1	5×10^1	5×10^1	5×10^1	4×10^1	3×10^1
Mn	6.8×10^2	2.5×10^2	4.6×10^3	9.8×10^2	1.3×10^3	5.1×10^2	4.1×10^2	1.5×10^3
Fe	3.4×10^4	1.5×10^4	3.5×10^4	3.7×10^4	2.5×10^4	3.3×10^4	2.6×10^4	2.4×10^4
Ni	20	17	88	20	36	20	12	22
Cu	25	50	99	5.7×10^2	5.3×10^2	1.7×10^2	71	3.4×10^2
Zn	4.6×10^2	1.3×10^2	1.9×10^3	2.0×10^2	1.5×10^3	1.1×10^2	1.5×10^2	1.7×10^3
As	<5	35	1.8×10^2	<5	2.9×10^2	<5	<5	34
Sr	1.5×10^2	2.5×10^2	1.2×10^2	2.9×10^2	2.1×10^2	90	1.3×10^2	2.3×10^2
Zr	2×10^2	2×10^2	5×10^1	1×10^2	2×10^2	1×10^2	1×10^2	1×10^2
Mo	<2.5	<2.5	10	<2.5	9	<2.5	<2.5	3
Ag	<1	3.0	3.5	2.8	3.4	1.6	1.2	2.6
Cd	<1	<1	4	<1	35	<1	<1	23
Au	<0.01	<0.01	<0.01	<0.01	<0.01	<0.01	<0.01	<0.01
Pb	<10	<10	9×10^1	2×10^1	1×10^3	<10	4×10^1	9×10^1
Th	<5	6	<5	<5	<5	6	<5	<5
U	<5	<5	<5	<5	<5	<5	<5	<5

*ss = very fine grained sandstone

**sts = coarse siltstone

Table B3. Geochemical data on unmineralized sandstone and siltstone of the Hazel Formation (values in ppm by weight).

	Mohawk mine		Marvin-Judson prospect	Hazel mine	Pecos mine
Element	**red ss* >100 ft from vein Sample No. 81-165**	**red ss within 10 ft of vein Sample No. 81-166**	**red ss 10 ft from vein Sample No. 82-12**	**red sts** within 10 ft of vein Sample No. 82-7**	**red ss >30 ft from vein Sample No. 82-30**
Li	33	41	10	16	26
Be	2.3	2.2	2.2	2.9	2.3
Na	2.7 x 10^{4}	2.4 x 10^{4}	2.1 x 10^{4}	1.7 x 10^{4}	2.6 x 10^{4}
Mg	1.1 x 10^{4}	1.2 x 10^{4}	9.3 x 10^{3}	2.3 x 10^{4}	1.3 x 10^{4}
Al	6.3 x 10^{4}	6.3 x 10^{4}	6.8 x 10^{4}	7.0 x 10^{4}	6.4 x 10^{4}
P	5.9 x 10^{2}	5.1 x 10^{2}	5.2 x 10^{2}	5.9 x 10^{2}	5.8 x 10^{2}
K	2.1 x 10^{4}	2.2 x 10^{4}	2.1 x 10^{4}	2.5 x 10^{4}	2.0 x 10^{4}
Ca	3.2 x 10^{4}	2.5 x 10^{4}	1.7 x 10^{4}	3.5 x 10^{4}	2.3 x 10^{4}
Ti	5.2 x 10^{3}	4.1 x 10^{3}	4.6 x 10^{3}	5.5 x 10^{3}	5.1 x 10^{3}
V	1.1 x 10^{2}	83	80	81	84
Cr	5 x 10^{1}	5 x 10^{1}	5 x 10^{1}	5 x 10^{1}	3 x 10^{1}
Mn	5.5 x 10^{2}	4.8 x 10^{2}	4.5 x 10^{2}	1.2 x 10^{3}	5.2 x 10^{2}
Fe	3.4 x 10^{4}	2.9 x 10^{4}	3.1 x 10^{4}	3.8 x 10^{4}	3.3 x 10^{4}
Ni	20	20	16	29	21
Cu	33	15	10	22	11
Zn	59	65	66	2.0 x 10^{2}	99
As	<5	<5	<5	14	9
Sr	86	1.2 x 10^{2}	1.2 x 10^{2}	1.8 x 10^{2}	1.1 x 10^{2}
Zr	2 x 10^{1}	1 x 10^{2}	1 x 10^{2}	2 x 10^{2}	2 x 10^{2}
Mo	<2.5	<2.5	<2.5	<2.5	<2.5
Ag	<1	<1	<1	<1	<1
Cd	<1	<1	<1	≤1	<1
Au	<0.01	<0.01	<0.01	<0.01	<0.01
Pb	<10	<10	<10	<10	<10
Th	<5	<5	7	<5	7
U	<5	<5	<5	<5	<5

*ss = very fine grained sandstone

**sts = coarse siltstone

Table B4. Geochemical data on other types of copper ore from the Hazel and Allamoore Formations (values in ppm by weight).

	Blackshaft mine			Sancho Panza mine	Anaconda No. 2 prospect
	fault gauge with 14-Å chlorite, quartz, sericite, and malachite	tuffaceous sedimentary rock containing quartz, calcite, graphite, digenite, djurleite, covellite, and malachite		tuffaceous sedimentary rock with quartz, feldspar, calcite, chlorite, malachite, and limonite	amygdaloidal basalt with malachite
Element	Sample No. 82-13	Sample No. 82-18	Sample No. 82-21	Sample No. 81-168	Sample No. 81-172
Li	51	85	1.3 x 10^{2}	1.2 x 10^{2}	41
Be	1.1	1.6	2.5	2.4	3.5
Na	1.6 x 10^{3}	5.3 x 10^{3}	2.6 x 10^{2}	2.4 x 10^{3}	3.3 x 10^{4}
Mg	2.2 x 10^{4}	6.4 x 10^{4}	7.1 x 10^{4}	7.7 x 10^{4}	2.3 x 10^{4}
Al	4.4 x 10^{4}	5.6 x 10^{4}	3.9 x 10^{4}	6.2 x 10^{4}	7.5 x 10^{4}
P	9.7 x 10^{2}	8.3 x 10^{2}	1.05 x 10^{3}	1.1 x 10^{3}	1.1 x 10^{3}
K	4.6 x 10^{3}	3.2 x 10^{3}	9.8 x 10^{2}	2.4 x 10^{3}	8.2 x 10^{3}
Ca	>4 x 10^{4}	2.4 x 10^{4}	>4 x 10^{4}	1.9 x 10^{4}	4.0 x 10^{4}
Ti	1.4 x 10^{3}	1.7 x 10^{3}	1.1 x 10^{3}	2.8 x 10^{3}	8.0 x 10^{3}
V	54	60	56	1.1 x 10^{2}	1.8 x 10^{2}
Cr	4 x 10^{1}	2 x 10^{1}	2 x 10^{1}	2 x 10^{1}	7 x 10^{1}
Mn	7.0 x 10^{2}	4.4 x 10^{2}	5.0 x 10^{2}	2.3 x 10^{2}	9.4 x 10^{2}
Fe	8.4 x 10^{3}	1.3 x 10^{4}	1.8 x 10^{4}	2.2 x 10^{4}	7.5 x 10^{4}
Ni	41	15	20	30	70
Cu	2.7 x 10^{4}	1.7 x 10^{4}	2.2 x 10^{4}	2.0 x 10^{4}	2.7 x 10^{4}
Zn	2.0 x 10^{3}	1.4 x 10^{2}	5.0 x 10^{2}	1.8 x 10^{2}	2.4 x 10^{2}
As	<5	<5	12	<5	<5
Sr	72	20	19	43	1.1 x 10^{3}
Zr	8 x 10^{1}	2 x 10^{1}	2 x 10^{1}	1 x 10^{2}	3 x 10^{1}
Mo	3	<2.5	3	<2.5	<2.5
Ag	15	14	3.7	7.3	8.4
Cd	2	<1	68	<1	2
Au	<0.01	<0.01	<0.01	<0.01	<0.01
Pb	<10	<10	1.5 x 10^{3}	<10	<10
Th	<5	<5	<5	<5	<5
U	<5	<5	<5	<5	<5

TABLE B5. Geochemical data on mineralized Yucca sandstones and conglomerates, Indio Mountains (values in ppm by weight).

	Purple Sage mine				Rossman prospect	Finlay prospect
	conglomerate samples with varying amounts of quartz, calcite, galena, malachite, chrysocolla, covellite, cerussite, wulfenite, mimetite, vanadinite, mottramite, hemimorphite, goethite, and manganese oxide				sandstone with quartz, barite, malachite, chalcocite, pyrite, and chalcopyrite	barite-jarosite-group mineral vein
Element	**Sample No.81-53**	**Sample No. 81-53hc**	**Sample No. 81-53lc**	**Sample No. 81-116c**	**Sample No. 82-63**	**Sample No. 82-37**
Li	17	n.a.*	n.a.	17	2	8
Be	<0.5	n.a.	n.a.	<0.5	<0.25	1.6
Na	5.5×10^2	7.2×10^2	1.7×10^2	4.9×10^2	2.0×10^2	4.3×10^2
Mg	5.2×10^2	3.5×10^2	6.8×10^2	4.8×10^2	4.0×10^2	1.8×10^2
Al	6.1×10^3	4.0×10^3	4.0×10^3	5.5×10^3	5.8×10^3	1.2×10^3
P	3.2×10^2	n.a.	n.a.	4.2×10^2	58	72
K	2.5×10^3	1.5×10^3	1.7×10^3	2.1×10^3	2.5×10^3	7.4×10^2
Ca	1.1×10^3	7.0×10^2	5.0×10^3	1.3×10^3	1.7×10^2	8.8×10^2
Ti	3.4×10^2	n.a.	n.a.	3.5×10^2	2.8×10^2	3.0×10^2
V	2.1×10^3	n.a.	n.a.	5.1×10^3	15	4
Cr	5×10^1	n.a.	n.a.	1×10^2	5	2
Mn	18	n.a.	n.a.	73	5	13
Fe	2.6×10^3	1.9×10^4	2.7×10^3	3.9×10^3	1.3×10^3	7.2×10^4
Ni	8	n.a.	n.a.	12	3	7
Cu	2.5×10^3	2.2×10^4	1.4×10^3	5.6×10^3	2.1×10^4	1.2×10^2
Zn	1.5×10^4	1.9×10^3	2.1×10^4	4.9×10^3	1.6×10^2	1.4×10^3
As	4.0×10^2	n.a.	n.a.	1.0×10^3	70	2.8×10^4
Sr	1.5×10^2	n.a.	n.a.	2.1×10^2	1.8×10^2	9
Zr	9	n.a.	n.a.	8	10	10
Mo	3.0×10^2	8.0×10^2	40	8.0×10^2	13	7
Ag	16	45	18	40	45	84
Cd	24	n.a.	n.a.	40	<1	17
Au	n.a.	n.a.	n.a.	n.a.	<0.01	<0.01
Pb	3.8×10^4	4.4×10^4	4.6×10^4	2.4×10^3	<10	2.7×10^3
Th	<10	n.a.	n.a.	<10	<5	<5
U	<10	n.a.	n.a.	20	<5	<5

*n.a. = not analyzed

TABLE B6. Geochemical data on weakly mineralized Yucca sandstones, Indio Mountains (values in ppm by weight).

	Finlay prospect, medium-grained sandstones		
	next to vein	**2 ft from vein**	**95 ft from vein***
Element	**Sample No. 82-39**	**Sample No. 82-40**	**Sample No. 82-43**
Li	13	3	3
Be	0.7	< 0.25	< 0.25
Na	2.1×10^2	1.1×10^2	1.6×10^2
Mg	9.6×10^3	5.6×10^2	2.7×10^2
Al	2.1×10^4	6.5×10^3	9.2×10^3
P	1.1×10^2	85	76
K	8.1×10^3	2.6×10^3	3.5×10^3
Ca	$> 4 \times 10^4$	2.3×10^4	2.8×10^2
Ti	9.8×10^2	2.8×10^2	4.2×10^2
V	33	8	16
Cr	1×10^1	4	6
Mn	6.2×10^2	7	1.3×10^2
Fe	2.3×10^4	4.7×10^3	5.9×10^3
Ni	12	6	6
Cu	11	1.1×10^2	11
Zn	5.8×10^2	47	16
As	3.1×10^2	2.3×10^2	1.6×10^2
Sr	1.1×10^2	98	25
Zr	10	10	10
Mo	25	< 2.5	≤ 3
Ag	< 1	< 1	< 1
Cd	3	< 1	< 1
Au	< 0.01	< 0.01	< 0.01
Pb	8.0×10^2	< 10	2.3×10^2
Th	< 5	< 5	< 5
U	< 5	< 5	< 5

*An unrecognized vein may be closer to this sample than is the Finlay vein.

TABLE B7. Geochemical data on unmineralized Yucca sandstones, Indio Mountains, and on average sandstone (values in ppm by weight).

	Rossman prospect		Finlay prospect		
	medium-grained white sandstone overlying ore-bearing strata	fine-grained red sandstone between veins	medium-grained sandstone 19 ft from vein	medium-grained sandstone 47 ft from vein	
Element	Sample No. 82-61	Sample No. 82-62	Sample No. 82-41	Sample No. 82-42	Average sandstone[1]
Li	8	6	2	4	15
Be	<0.25	0.3	<0.25	<0.25	0.5
Na	1.5×10^2	1.6×10^2	1.3×10^2	1.3×10^2	4.2×10^3
Mg	6.7×10^2	1.3×10^3	2.3×10^2	2.3×10^2	1.1×10^4
Al	9.5×10^3	1.3×10^4	8.3×10^3	7.3×10^3	4.2×10^4
P	85	58	1.9×10^2	87	1.7×10^2
K	4.3×10^3	5.9×10^3	3.1×10^3	2.9×10^3	1.07×10^4
Ca	6.2×10^3	3.3×10^4	2.1×10^2	3.8×10^2	3.1×10^4
Ti	4.2×10^2	1.2×10^3	3.9×10^2	3.3×10^2	3.4×10^3
V	19	20	14	11	20
Cr	7	1×10^1	5	5	35
Mn	35	4.7×10^2	1.2×10^2	2.6×10^2	5.0×10^2
Fe	2.3×10^3	7.8×10^3	5.7×10^3	6.4×10^3	9.8×10^3
Ni	4	6	4	5	2
Cu	82	2	7	9	10
Zn	4	7	8	8	40
As	12	7	26	50	1.2
Sr	28	55	18	23	20
Zr	1×10^1	2×10^1	3×10^1	8	2.2×10^2
Mo	≤ 2.5	< 2.5	≤ 3	≤ 3	0.2
Ag	<1	<1	<1	<1	0.25
Cd	<1	<1	<1	<1	0.05
Au	< 0.01	< 0.01	< 0.01	< 0.01	0.005
Pb	<10	<10	<10	2×10^1	10
Th	<5	<5	<5	<5	5.5
U	<5	<5	<5	<5	1.7

[1]from Wedepohl (1969); Rose and others (1979)

TABLE B8. Geochemical data on limestone and calcite in the Yucca Formation, Indio Mountains (values in ppm by weight).

	Finlay prospect		
	mineralized block of thin limestone bed in fault zone	**unmineralized thin limestone bed**	**north of Road prospect—calcite vein along fault**
Element	**Sample No. 82-36**	**Sample No. 82-38**	**Sample No. 82-1**
Li	<1	2	<1
Be	0.3	<0.25	<0.25
Na	1.7×10^2	1.7×10^2	1.2×10^2
Mg	3.0×10^3	5.4×10^3	4.0×10^3
Al	8.6×10^3	6.3×10^3	5.0×10^2
P	1.3×10^2	98	81
K	3.6×10^3	2.4×10^3	1.3×10^2
Ca	$>4 \times 10^4$	$>4 \times 10^4$	$>4 \times 10^4$
Ti	5.2×10^2	4.4×10^2	24
V	13	9	7
Cr	6	4	1
Mn	1.2×10^3	1.4×10^3	1.2×10^3
Fe	1.3×10^4	4.2×10^3	1.0×10^3
Ni	6	<2.5	<2.5
Cu	1.2×10^2	30	10
Zn	2.9×10^3	34	38
As	9.4×10^2	23	<5
Sr	1.5×10^2	1.2×10^2	1.2×10^2
Zr	14	6	<1
Mo	21	≤ 2.5	< 2.5
Ag	20	<1	n.a.*
Cd	2.1×10^2	<1	<1
Au	<0.01	<0.01	n.a.
Pb	4.0×10^3	11	1.5×10^2
Th	<5	<5	<5
U	<5	<5	<5

*n.a. = not analyzed